Studies in Fuzziness and Soft Computing

Volume 335

About this Series

The series "Studies in Fuzziness and Soft Computing" contains publications on various topics in the area of soft computing, which include fuzzy sets, rough sets, neural networks, evolutionary computation, probabilistic and evidential reasoning, multi-valued logic, and related fields. The publications within "Studies in Fuzziness and Soft Computing" are primarily monographs and edited volumes. They cover significant recent developments in the field, both of a foundational and applicable character. An important feature of the series is its short publication time and world-wide distribution. This permits a rapid and broad dissemination of research results.

More information about this series at http://www.springer.com/series/2941

Mikael Collan · Mario Fedrizzi
Janusz Kacprzyk
Editors

Fuzzy Technology

Present Applications and Future Challenges

Springer

Editors
Mikael Collan
School of Business and Management
Lappeenranta University of Technology
Lappeenranta
Finland

Janusz Kacprzyk
Polish Academy of Sciences
Warsaw
Poland

Mario Fedrizzi
Universita di Trento
Trento
Italy

ISSN 1434-9922 ISSN 1860-0808 (electronic)
Studies in Fuzziness and Soft Computing
ISBN 978-3-319-26984-9 ISBN 978-3-319-26986-3 (eBook)
DOI 10.1007/978-3-319-26986-3

Library of Congress Control Number: 2015955600

Springer Cham Heidelberg New York Dordrecht London

Printed on acid-free paper

Springer International Publishing AG Switzerland is part of Springer Science+Business Media
(www.springer.com)

To Professor Christer Carlsson

This book is a token of appreciation to Prof. Christer Carlsson for his long-time service to the fuzzy logic community, for his support and inspiration that have shaped interests and careers of many people, for his integrity and quest for scientific excellence, and for his relentless and successful promotion of the use of fuzzy logic in business, technology, and everywhere. Christer Carlsson is one of the most prominent grand personalities in decision-making research in Finland today, what makes him special is the fact that he has been able to combine cutting-edge research with industrial collaboration to the benefit

of the researchers and the companies involved. In many research projects supported by the Finnish Funding Agency for Innovation (TEKES), he and his research team have solved many real-world problems by creating new answers for old and new problems.

The variety of problems Christer has been involved in solving is impressive: ranging from MCDM problems in the early days of the WWW and using hypertext documents in company intranets, to project evaluation by using real option analysis and the creation of the first fuzzy variants for real option valuation, to using soft computing methods in decision analytics.

Application areas have been diverse, many times including the management of large international corporations' portfolios of patents, R&D projects, prospective acquisition targets, etc., in a future-oriented way, by integrating imprecise and scenario-based numerical information with managerial insight and tacit knowledge—and all this, in the age before "big data," and before the ongoing revolution of data-based management.

Christer has been strongly involved in research in the field of information systems and especially in the research of mobile services from the very beginning of their emergence. Today, mobile services, soft computing, and decision-making are topics that have converging paths; it is through mobile devices that information is being collected in many ways and this information can be used in decision-making. It is not by coincidence Christer and his research group have a strong interest and experience in this convergence.

The hallmark of a visionary is the immense amount of time and energy spent in trying to seriously understand what is going to happen in the future and being able to put the pieces together in an innovative way that adds value. It is apparent to anyone that knows Christer that he has never spared the effort in finding out what is new, what is hot, and what will be the next big thing—it is this persistent curiosity that drives great scientists and is what this book celebrates together with Christer.

Preface

This volume is a result of a special project the purpose of which was twofold. First of all, we wished to provide a bird's eye view of some novel directions in the broadly perceived "fuzzy logic," starting from more philosophical and foundational considerations, through a bunch of promising models for the analysis of data, decision-making, and systems modeling, to a logical consequence in view of the very topic covered by this volume, reflected in its title, that is, an account of some successful experiences in using fuzzy technology in practice, in business, and in technology. Second, from a more personal perspective, we wished this volume to be a token of appreciation to Prof. Christer Carlsson, from the Institute for Advanced Management Systems Research (IAMSR), Åbo Akademi University in Åbo (Turku) in Finland; a very special person, a friend and a peer, to the entire fuzzy logic community, who has managed for so many years to combine top-level theoretical research with applications, and finally implementations of various fuzzy logic based models in business and technology. Of course, before entering the field of fuzzy logic he was a prominent representative of the operations research community, notably that in multiple-criteria decision-making (MCDM).

What concerns the last two decades of his activities is that he has been a driving force behind the foundation and an unprecedented growth of his large research group, IAMSR, a rare research institution not only in Europe, but all over the world, which has been able to, for many years, continuously collect funds from research and development projects with business and industry. This should be considered as a sign of Prof. Carlsson's vision and far-sighted ideas on the fact that a close collaboration with practice is crucial for the academic world. After many years of personal success, this vision is today confirmed by the policies of both the European Union and other governments around the world.

Christer was a visionary and a proponent of modern applications of fuzzy technology, and—what has been rare and difficult—he has always been, and is, one of those not so numerous people in our academic community who have been able to secure substantial industrial grants for large research groups. It is worth mentioning that in his works he has always been able to find a synergistic combination of high scientific level with real-world applications.

Over many years he has been an active member of the International Fuzzy Systems Association (IFSA), its long-time Council member, Treasurer, and finally President. His devotion and vision have certainly contributed to an extraordinary success of IFSA and proliferation of fuzzy logic all over the world.

We are sure that this volume will be interesting to many people who deeply feel, as Christer has always felt, that a synergistic combination of high-level theory with real applications is not only possible, but also is necessary, and can certainly be successful.

Following what has been said, we start our volume with Part I: Foundations, in which some papers from a more general to some extent philosophical type are included, which are important, because they clarify many aspects and issues that are relevant both for the foundations and soundness of many applications. Enric Trillas and Rudolf Seising in their "Turning Around the Ideas of Meaning and Complement" consider an important problem that has been with fuzzy sets/logic since its inception, namely the fact that though fuzzy sets have been viewed as specifying the meaning of some linguistic terms. There has not been many works on a deeper analysis of how the meaning should be meant so that those linguistic terms could be subject of a scientific inquiry. In the paper the authors follow the path rooted in the Wittgenstein's 'identification' of meaning and linguistic use, in which meaningful predicates can be seen as those represented by quantities, whose measures are the membership functions. Unfortunately, fuzzy set theory is still missing a complete study on the meaning of connectives that, in fuzzy logic, are not universal, as they are in classical logic, but are context-dependent and purpose-driven. This issue is discussed in depth, as well as issues related to a proper definition of the negation. Deeper semantic analyses for concepts like negation or antonym in general are usually not addressed, since they are often just applied through the membership functions. In this paper, the idea of a linguistic concept of 'complement,' either meant as 'not' or is 'opposite' is analyzed. For a person working in the theory of fuzzy logic, some ideas of the practice of natural language are also included.

Jorma K. Mattila in "A Note on Fuzzy-Valued Inference" considers fuzzy-valued inference, which is analyzed by using concepts and properties of a theory of fuzzy-valued associative Kleene algebra. To be more specific, it is shown that the tools and techniques presented can be used to develop a fuzzy screening system as introduced by Yager.

Vesa A. Niskanen in "A Concept Map Approach to Approximate Reasoning with Fuzzy Extended Logic" discusses the important problem of how to develop computer models of approximate reasoning, notably in terms of modus ponens and modus tolens. He proposes the use of Zadeh's fuzzy extended logic and the idea of a concept map which has been widely used in many areas. An application in statistics is shown.

In Part II: Data Analysis, Decision-Making and Systems Modeling, we included papers that have presented more specific models for solving various problems of relevance. First, some problems related to broadly perceived data analysis are dealt with. Christian Borgelt and David Picado-Muíno in "Significant Frequent Item Sets via Pattern Spectrum Filtering" deal with the important problem of the frequent item set mining. Unfortunately, set mining is often plagued by the fact that the sets of frequent items can be very large, in some cases the size of the output can even exceed the size of the set of data involved. The authors propose new extensions to some statistical approaches that produce only significant frequent item sets (or association rules derived from them), which combines data randomization with a so-called pattern spectrum filtering.

Adriano S. Koshiyama, Marley M.B.R. Vellasco, and Ricardo Tanscheit in "A Novel Genetic Fuzzy System for Regression Problems" consider a regression problem that is equivalent to finding a model that relates the behavior of an output or response variable to a given set of input or explanatory variables, and has a universal importance and applicability. To meet the requirements of linguistic interpretability and reasonable accuracy, the authors present a novel genetic fuzzy system (GFS), called genetic programming fuzzy inference system for regression problems (GPFIS-Regress). The system utilizes multi-gene genetic programming to build the premises of fuzzy rules, with various t-norms, negation, and linguistic hedge operators. The system is tested on some relevant benchmark examples and the results are promising.

Silvia Bortot, Mario Fedrizzi, Michele Fedrizzi, and Ricardo Alberto Marques Pereira in "A Multidistance Approach to Consensus Modeling" investigate the relationship between the soft measure of collective dissensus and the multidistance approach to consensus evaluation previously introduced by them. They propose a new approach in which a particular type of sum-based multidistance used as a measure of dissensus is defined. This multidistance is characterized by the application of a subadditive filtering function the effect of which is that small values of distances are emphasized and large ones are attenuated. In an example, a comparison of the new dissensus measure with the OWA-based multidistance obtained assuming that the weights are linearly decreasing with respect to increasing distance values is performed.

Janusz Kacprzyk, Dominika Gołuńska, and Sławomir Zadrożny in "A Consensus Reaching Support System Based on the Concepts of an Ideal and Anti-Ideal Agent and Option" an extension of a series of previous works on a moderator run consensus reaching process, and its related group decision-making process in a small group of autonomous decision-makers (agents) is presented. The approach proposed is based on fuzzy preferences, fuzzy majority represented as linguistic quantifiers, and fuzzy majority based soft measure of the consensus, and the emphasis is on the running of a consensus reaching process via a moderator. To help the moderator to run the process, additional higher-level information, notably in the form of linguistic data summaries is used. Specifically, a new concept of an ideal and anti-ideal point and its related TOPSIS method based approach, is proposed.

Elid Rubio, Oscar Castillo, and Patricia Melin in "Interval Type-2 Fuzzy System Design Based on the Interval Type-2 Fuzzy C-Means Algorithm" use the interval type-2 fuzzy C-means (IT2FCM) algorithm to develop an interval type-2 fuzzy inference systems, using the centroids and fuzzy membership matrices for the lower and upper bound of the intervals obtained by the IT2FCM algorithm, in each data clustering step that occurs in this algorithm. Using these tools and techniques, the Mamdani, and Sugeno fuzzy inference systems for classification of data sets and time series prediction are developed.

Pavel Holeček, Jana Talašová, and Jan Stoklasa in "Multiple-Criteria Evaluation in the Fuzzy Environment Using the FuzzME Software" describe a software tool for fuzzy multiple-criteria evaluation called "FuzzME," and present how to apply the software for solving a broad range of fuzzy MCDM problems. The mathematical foundation on which the FuzzME software is built is also briefly described. An interesting feature is that for the aggregation of partial scores, various aggregation methods can be employed, notably fuzzy weighted average, fuzzy OWA operator, and fuzzified WOWA operator. In the case of interacting criteria, the fuzzified discrete Choquet integral, and also an aggregation function described by fuzzy rule base, defined by the experts, can be used.

Part III: Fuzzy Logic in Business and Industrial Practice concerns very important works related to the practical use of broadly perceived intelligent, notably fuzzy, technologies, and experience gained. Mikael Collan and Pasi Luukka in "Strategic R&D Project Analysis: Keeping It Simple and Smart" deal with strategic R&D projects that are the core of virtually all high-scale undertakings in many areas of human activities. They consider the situation, where forward-looking analysis is required and assume that the decision-makers face structural uncertainty. Detailed, precise information is very often, if not always, not available in the considered situation. This implies a need for robust management systems that are capable of handling imprecise and uncertain information, yet simple and comprehensive enough for managerial use. The authors propose a set of new weighted averaging operators that are able to consider interaction between variables and an approach that is based on scorecards in which triangular fuzzy numbers are used. This gives a simple, easy to understand, easy to visualize, low-cost, multi-expert analysis tool for strategic R&D projects that can be implemented on a laptop using spreadsheet software.

József Mezei and Matteo Brunelli in "Decision Analytics and Soft Computing with Industrial Partners: A Personal Retrospective" consider methods in decision analytics, which become essential tools to process an increasing amount of collected data, while being capable of representing and utilizing the tacit knowledge of experts. This boils down to a need from companies of methods that can make use of imprecise information to deliver insights in real time. The authors provide a summary of three closely related research projects within the area of knowledge mobilization, and provide solutions for typical business analytical problems, originating mainly from the process industry. They use fuzzy ontologies, represented as fuzzy relations. By analyzing the similarities among the three presented cases, they discuss the main lessons learnt and provide some advice as what should be considered in future applications of soft computing in industrial applications.

J.M. Sánchez-Lozano, M.S. García-Cascales, M.T. Lamata, and J.L. Verdegay in "Spatial Analysis Using GIS for Obtaining Optimal Locations for Solar Farms—A Case Study: The Northwest of the Region of Murcia" consider an important problem of making decisions as to where to locate a photovoltaic solar farm, to provide the energy generated into the grid. They consider, first, the legislative factors involving a large number of restrictions with regards to protected areas, streams, watercourses, etc., and then some criteria exemplified by the proximity to power lines, slope, solar irradiation, etc., according to which an evaluation of the suitability of the areas will be made. They use spatial visualization tools such as geographic information systems (GIS). The main purpose is to show how the aggregation of GIS with decision procedures in the field of renewable energy can solve complex location problems. An example of determining suitable locations for photovoltaic solar farms in the northwest region of Murcia in Spain is shown.

We wish to thank all the contributors to this volume. We hope that their papers, which constitute a synergistic combination of foundational works, new data analysis tools and techniques, new decision-theoretic and -analytic models, new methodologies and techniques for system modeling, and papers on relevant real-world implementations, combined with an account of experience gained from various real and successful practical projects will be interesting and useful for a large audience.

We also wish to thank Dr. Tom Ditzinger, Dr. Leontina di Cecco, and Mr. Holger Schaepe from Springer for their dedication and help to implement and finish this publication project on time, maintaining the highest publication standards.

August 2015

Mikael Collan
Mario Fedrizzi
Janusz Kacprzyk

Contents

Part I
Foundations

Turning Around the Ideas of 'Meaning' and 'Complement'

Enric Trillas and Rudolf Seising

Abstract Since the inception of fuzzy sets in 1965, Lotfi A. Zadeh viewed them by specifying the meaning of their linguistic labels in a given universe of discourse. Anyway, it lacked to clarify the same concept of meaning in a way able to separate those words that can be submitted to scientific scrutiny from those that cannot, as well as to show which it is, into meaning, the actual role of the membership function of a fuzzy set. Once rooted in the Wittgenstein's 'identification' of meaning and linguistic use, meaningful predicates can be seen as those represented by quantities whose measures are the membership functions. Closed such question, it still lacks to completely study the meaning of connectives that, in fuzzy logic, are not universal as they are in classical logic, but context-dependent and purpose-driven; hence and like fuzzy sets, should be carefully designed at each case. When fuzzy logic was theoretically introduced, from the mid-sixties to the last eighties of the XX century, crisp deductive logic ideas were prevalent over those of conjectural ordinary reasoning to represent, for instance, what is 'not properly covered under a linguistic label' and only the concept of pseudo-complement was considered, even if antonyms are essential for linguistic variables, a basic tool in the applications of fuzzy logic. Fifty years later, when fuzzy logic comes to be surpassed by Zadeh's 'Computing with Words', perhaps some thoughts at the respect could be suitable towards its applicability to represent complex linguistic statements in common sense reasoning. This paper just tries to reflect on one of the subjects fuzzy logic perhaps manages in a too simplistic way, almost uniquely by means of the strong negation 1-id, and from an analogous point of view to that of logic that not always is close to ordinary reasoning. It refers to a general concept of 'complement', a concept that tries to 'collectivize' all that does not properly lay under a given concept but should not to be forgot since it completes what is taken into account. In the, say traditional treatment of set's complement, it lacks the consideration of what actually happens in language, the true background of what fuzzy logic tries to represent, where 'opposites' play a role that is, at least, weakly but equal or more important that that played by 'not'. Fuzzy logic's praxis just

To Professor Christer Carlsson.

E. Trillas (✉) · R. Seising
European Centre for Soft Computing, Mieres, Asturias, Spain
e-mail: enric.trillas@softcomputing.es

M. Collan et al. (eds.), *Fuzzy Technology*, Studies in Fuzziness and Soft Computing 335, DOI 10.1007/978-3-319-26986-3_1

considers both negation and antonym through their membership functions, avoids a deep semantic analysis of both concepts, has not criteria to decide when membership functions should, or should not, be functionally expressed from the membership function of the initial linguistic label, as well as how it can be done with the known models of negation functions. The process followed by fuzzy logic's practitioners for designing opposites and negation is often too lose and quickly done for well adjusting representation to what is been represented. The point of view taken in this paper is not mathematical in nature, but is just a first trial to reflect on what can surround a linguistic concept of 'complement', of what is either 'not', or is 'opposite', to what is qualified by a predicate. The paper only tries to open the eyes of theoreticians towards the true ground on which fuzzy logic is anchored, the natural language's practice; fuzzy logic cannot be neither a mathematical subject, nor one of just a computational interest, but a discipline of the imprecise similar to an experimental science.

1 Introduction

1.1

If what can be qualified as 'fuzzy' is seen as a matter of degree, of gradation, the practical use of fuzzy sets for computationally modeling fuzzy systems is actually a matter of design, depending upon the context in which the linguistic labels currently apply to, the purpose for its use, as well as the mathematical and computational armamentarium available for its representation. The concept of a fuzzy set is linked with that of measurable or gradable meaning, and its genetic roots lie in the linguistic phenomenon of creating collectives named by a precise or imprecise predicate, like the origin of thermodynamics lies in heat's transmission, a phenomenon that is based on a collective of molecules and their velocities. Fuzzy sets with the same linguistic label are nothing else than the states in which collectives manifest themselves with each contextual use of the linguistic label; particular fuzzy sets are situation-dependent.

The representation of collectives by fuzzy sets cannot ignore how the use of linguistic labels is learned and acquired in natural language; for instance, the meaning of 'John is tall' is difficult to be captured without simultaneously capturing that of 'Peter is short', or the use of 'less than five' without that of 'more than six'. Pairs like tall/short, less/more, etc., should be taken into account for designing a membership function for either tall or short, etc. Fuzzy logic is indeed rooted in the natural phenomena of language and ordinary reasoning; hence it is not static at all, and should be closer to experimental science than to what is understood by logic; in particular, because logic is essentially discrete and shows no flexibility enough for taking into account the conspicuous flexibility of language. Logic's main worries turn around formal deductive reasoning, and those of science with natural phenomena; natural language and ordinary reasoning are, at the end, natural phenomena deeply rooted in the functioning of the human brain. It is difficult to imagine how thermodynamics, for instance, could evolve from the old XVIII and XIX centuries problems on heat's transmission only by means of applying to them a purely logical treatment. Of course a 'logical control' of the involved kinds of reasoning cannot be avoided; it is perhaps thanks to this control that Fourier's Transform is known.

Fuzzy set theory mainly deals with those linguistic concepts that are imprecise and/or uncertain in a non-random or repetitive way, and most of which are linguistic terms, that is, can be found in the dictionary. Fuzzy logic would better deserve to be called the science of imprecision, and sooner or later it will need to escape from the logical habit of studying the meaning of large sentences through its elemental or atomic components as it is done in the case of classical/crisp logic; in ordinary life the meaning of these components is often captured after capturing the meaning of the full sentence through applying memory and ordinary reasoning to concepts that are, almost always, imprecise and uncertain. It will take some time to escape from such kind of atomism, and up to when the study of language can move to a true experimental science in which computationally-assisted controlled observation and mathematical models can play a basic role; towards a kind of physics of language and reasoning. For instance, the currently done study of antonyms in linguistics avoids taking into account the degrees in which the statements with these terms do hold, and this fact seriously limits a study needing to move from one to two dimensions, like Max Black tried to do a lot of years ago with his profile functions [1].

To cite something else, it is not the same negating an elemental statement 'x is P', than negating a large sentence containing elemental statements but involving linguistic hedges, connectives, quantifiers, etc., and if for some linguistic terms there exist an antonym, it is not still clear if it there are actually antonyms of large statements, as it is also not clear its inexistence for crisp linguistic terms. Thinking based on a lot of intentional experimentation waits to be done, but the pervasiveness of imprecision and non-repetitive uncertainty in language makes to think that a, perhaps renewed, fuzzy logic can play a relevant role in it in the way towards an experimental science of language and commonsense reasoning supplied with a measuring capability. Currently it is actually difficult to imagine some experimental science without more or less sophisticated systems of measuring what is involved in it.

1.2

Mathematics, Logic and philosophy of language are closely connected and in the 20th century it was Gottlob Frege who wrote the classic article "Über Sinn und Bedeutung" ("On Sense and Reference") [2]. "Sense" and "reference" denote in Frege's philosophy two aspects of the meaning of a term: its reference is the object to which the term refers, while its sense is the way that the term refers to the object. Frege also combined the subject of meaning with that of vagueness. In a lecture of 1891, he introduced concept as follows: "If '$x + 1$' is meaningless for all arguments x, then the function $x + 1 = 10$ has no value and no truth value either. Thus, the concept "that which when increased by 1 yields 10" would have no sharp boundaries. Accordingly, for functions the demand on sharp boundaries entails that they must have a value for every argument [3].[1] Few years later he published *Grundgesetze der Arithmetik (Foundations of Arithmetic)*, where he called for concepts with sharp boundaries, because otherwise we could break logical rules and, moreover, the conclusions we draw could be false (Frege, 1893–1903): "A definition of a concept (of a possible predicate) must be complete; it must unambiguously determine, as regards any object, whether or not it falls under the concept (whether or not the predicate is truly ascribable to it). [...] We may express this metaphorically as follows: the concept must have a sharp boundary" ([4], Sect. 56).[2] He also drew consequences if we would fail to follow this rule:

> To a concept without sharp boundary there would correspond an area that had not a sharp boundary-line all round, but in places just vaguely faded away into the background. This would not really be an area at all; and likewise a concept that is not sharply defined is wrongly termed a concept" ([4], Sect. 56).[3]

Frege's specification of vagueness as a particular phenomenon influenced other scholars, notably his British contemporary and counterpart the philosopher and mathematician Bertrand Russell, who wrote in his article 'Vagueness' in 1923 [5].

> Let us consider the various ways in which common words are vague, and let us begin with such a word as 'red'. It is perfectly obvious, since colours form a continuum, that there are shades of color concerning which we shall be in doubt whether to call them red or not, not because we are ignorant of the meaning of the word 'red', but because it is a word the extent of whose application is essentially doubtful. This, of course, is the answer to the old puzzle about the man who went bald. It is supposed that at first he was not bald, that he lost his hairs one-by-one, and that in the end he was bald; therefore, it is argued, there must have been one hair the loss of which converted him into a bald man. This, of course, is absurd. Baldness is a vague conception; some men are certainly bald, some are certainly not bald, while between them there are men of whom it is not true to say they must either be bald or not bald ([5], p. 85).

Russell reasoned "that all words are attributable without doubt over a certain area, but become questionable within a penumbra, outside which they are again certainly not attributable" ([5], p. 86 f). Then he generalized that words of pure logic also have no precise meanings, e.g. in classical logic the composed proposition 'p or q' is false only when p and q are false and true elsewhere. He went on to claim that the truth values "'true' and 'false' can only have a precise meaning when the symbols

---

[1]This is a mathematical verbalization of what is called the classical sorites paradox that can be traced back to the old Greek word $\sigma o \rho o \varsigma$ (for 'heap') used by Eubulid of Alexandria (4th century BC).

[2]"Eine Definition eines Begriffes (möglichen Prädikates) muss vollständig sein, sie muss für jeden Gegenstand unzweideutig bestimmen, ob er unter den Begriff falle (ob das Prädikat mit Wahrheit von ihm ausgesagt werden könne) oder nicht [...]. Man kann das bildlich so ausdrücken: der Begriff muss scharf begrenzt sein ..." ([3], p. 69).

[3]"Einem unscharf begrenzten Begriffe würde [wenn man sich Begriffe ihrem Umfang nach als Bezirke in der Ebene versinnlicht] ein Bezirk entsprechen, der nicht überall eine scharfe Grenzlinie hätte, sondern stellenweise ganz verschwimmend in die Umgebung überginge. Das wäre eigentlich gar kein Bezirk; und so wird ein unscharf definierter Begriff mit Unrecht Begriff genannt ([3], p. 70).

employed—words, perceptions, images ...—are themselves precise". As we have seen above, this is not possible in practice, so he concludes "that every proposition that can be framed in practice has a certain degree of vagueness; that is to say, there is not one definite fact necessary and sufficient for its truth, but certain region of possible facts, any one of which would make it true. And this region is itself ill-defined: we cannot assign to it a definite boundary". Russell emphasized that there is a difference between what we can imagine in theory and what we can observe with our senses in reality: "All traditional logic habitually assumes that precise symbols are being employed. It is therefore not applicable to this terrestrial life, but only to an imagined celestial existence" ([5] p. 88 f). He proposed the following definition of accurate representations: "One system of terms related in various ways is an accurate representation of another system of terms related in various other ways if there is a one-one relation of the terms of the one to the terms of the other, and likewise a one-one relation of the relations of the one to the relations of the other, such that, when two or more terms in the one system have a relation belonging to that system, the corresponding terms of the other system have the corresponding relation belonging to the other system". And in contrast to this, he stated that "a representation is vague when the relation of the representing system to the represented system is not one-one, but one-many" ([5], p. 89). He concluded that "Vagueness, clearly, is a matter of degree, depending upon the extent of the possible differences between different systems represented by the same representation. Accuracy, on the contrary, is an ideal limit" ([5], p. 90).

In his *Tractatus logico-philosophicus* Wittgenstein distinguished between meaningful, meaningless and nonsensical statements. The statements of logic, i.e. tautologies and contradictions are senseless, because "Tautologies and contradictions are not pictures of reality. They do not represent any possible situations. For the former admit all possible situations, and latter none" ([6], 4.462). By contrast, senseful propositions may "they range within the truth-conditions drawn by the propositions of logic. But the propositions of logic themselves are neither true nor false. [...] Other (non-logical) propositions can be senseless because they may apply to things that "cannot be represented, such as mathematics or the pictorial form itself of the pictures that do represent. These are, like tautologies and contradictions, literally sense-less, they have no sense'. [...] Nonsensical propositions cannot carry sense, e.g. 'Socrates is identical". "While some nonsensical propositions are blatantly so, others seem to be meaningful-and only analysis carried out in accordance with the picture theory can expose their nonsensicality. Since only what is 'in' the world can be described, anything that is 'higher' is excluded, including the notion of limit and the limit points themselves. Traditional metaphysics, and the propositions of ethics and aesthetics, which try to capture the world as a whole, are also excluded, as is the truth in solipsism, the very notion of a subject, for it is also not 'in' the world but at its limit" [1].

One of the best specialists on Frege's, Russell's and Wittgenstein's philosophical systems was Max Black who wrote *A Companion to Wittgenstein's Tractatus* in 1964. In 1937 Black published the article "Vagueness. An Exercise in Logical Analysis" where he defended the thesis of the measurability of linguistic vagueness by using

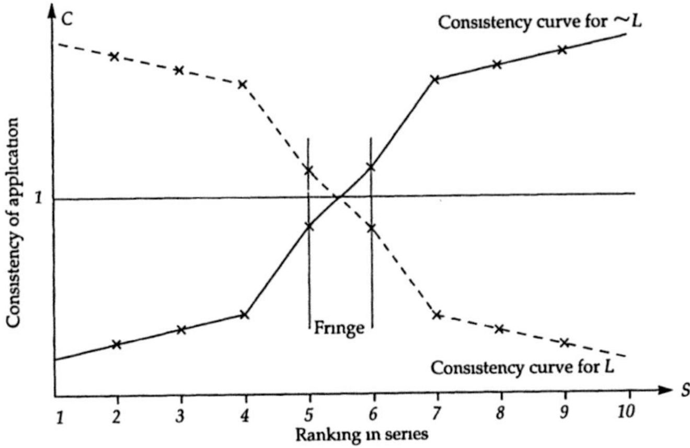

Fig. 1 Consistency of application of a typical vague symbol ([1], p. 443)

so-called consistency profiles. Black assumed that the vagueness of a word—as an element of a language—involves variations in its application by different users of that language. Furthermore he thought that these variations fulfill systematic and statistical rules when one symbol has to be discriminated from another. A speaker of a language has to decide whether to apply the word L or its negation $\neg L$ to an object x. "Such a situation arises, for instance, when an engine driver on a foggy night is trying to decide whether the light in the signal box is really a red or a green light" ([1], p. 442). He defined this discrimination of a symbol x with respect to a symbol L by DxL (and he wrote that we obtain $DxL = \mathrm{Dx}\neg L$ by definition). Then, he claimed that most speakers of a language and the same observer in most situations will determine that either L or $\neg L$ is used. In both cases, among competent observers there is a certain unanimity, a preponderance of correct decisions. For all DxL with the same x but not necessarily the same observer, m is the number of L uses and n the number of $\neg L$ uses. On this basis, Black stated the following definition: "We define the consistency of application of L to x as the limit to which the ratio m/n tends when the number of DxL and the number of observers increase indefinitely. [...] Since the consistency of the application, C, is clearly a function of both L and x, it can be written in the form $C\ (L, x)$" ([1], p. 442, Fig. 1).

Carl Gustav Hempel, who was associated with the Vienna Circle and immigrated in 1937 to the US, criticized Black's theory and pleaded for an extension of the research of vagueness from linguistics to a much broader field: "To what scientific discipline does the study of vagueness belong? No doubt, it is a task incumbent on a general theory of signs, or, to employ the term used by C.W. Morris, on semiotic". Referring to Charles Morris' *Foundation of the Theory of Signs*, published in the foregoing year [7], Hempel wrote that vagueness is 'strictly semiotical term, its determination requires reference to the symbols, to the users, and their designate' [8], p. 166. Thus, he claimed that 'complete statement on the vagueness of a symbol is of the following type: 'The vagueness of the term T, as applied, by the group G of persons, to the elements of the series S of objects, is v'. Thus, *vagueness* is what may be called a three-place semiotic relation which may assume different degrees (the values of v), or, in more technical terms: a strictly semiotic function of three arguments (T, G, S) being its arguments and v its value).

Differently to his opinion concerning the meaning of words and sentences in the *Tractatus* Wittgenstein considered in his late philosophy the user of a language important. In his *Philosophical Investigations* that Wittgenstein started writing in the 1930s, he let go the idea of a pure logical and therefore non-vague language. Thus, he came to the important result on what is meaning in his philosophy of language: "For a *large* class of cases—though not for all—in which we employ the word 'meaning' it can be defined thus: the meaning of a word is its use in the language" [9].

Our thesis in this book contribution is that meaning, vagueness and measurability are interlinked concepts. 'Measure what is measurable and make measurable what is not so' is a sentence attributed to Galileo. History of modern science shows the path from perceptions and observations to measurable quantities and therefore also from statements built up with perceptive predicates to statements built up with measurable predicates. In the following sections we will discuss the relationships of such statements and their meanings. We will argue that the concept of the meaning of a scientifically-fertile predicate lies in its measurability's character.

However, before we will go into details we will recall the connection of meaning and fuzziness in the early view of Lotfi Zadeh in the following subsection.

1.3

About 5 years after the appearance of "Fuzzy Sets" Lotfi Zadeh started studies on Fuzziness and Meaning. He wondered 'Can the fuzziness of meaning be treated quantitatively, at least in principle?' ([10], p. 160). In "Quantitative Fuzzy Semantics" he wrote: 'Few concepts are as basic to human thinking and yet as elusive of precise definition as the concept of 'meaning'. Innumerable papers and books in the fields of philosophy, psychology, and linguistics have dealt at length with the question of what is the meaning of meaning without coming up with any definitive answers'[4] ([10], p. 159]).

His studies on the 'fuzziness of meaning in a quantitative way' started as follows:

> Consider two spaces: (a) a universe of discourse, U, and (b) a set of terms, T, which play the roles of names of subsets of U. Let the generic elements of T and U be denoted by x and y, respectively'. Then he started to define the meaning $M(x)$ of a term x as a fuzzy subset of U characterized by a membership function $\mu(y \mid x)$ which is conditioned on x ([10], p. 164f]).

He gave an example: "Let U be the universe of objects which we can see. Let T be the set of terms *white, grey, green, blue, yellow, red, black*. Then each of these terms, e.g., *red*, may be regarded as a name for a fuzzy subset of elements of U which are red in color. Thus, the meaning of red, M (red), is a specified fuzzy subset of U" ([10], p. 164f).

[4]In a footnote he named the works of 12 known philosophers, linguists or cognitive scientists.

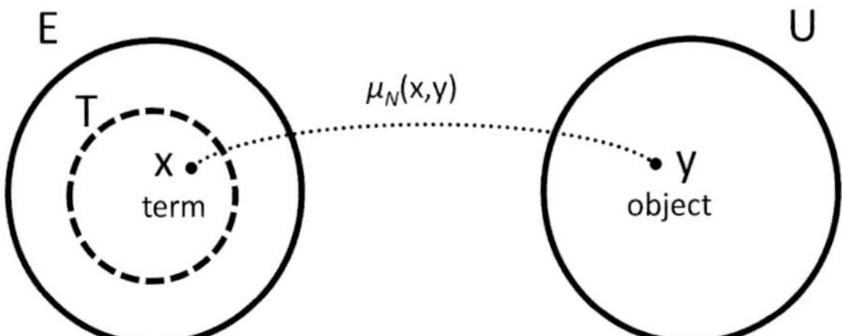

Fig. 2 The components of a fuzzy language: U = universe of discourse; T = term set; E = embedding set for T; N = naming relation from E to U; x = term; y = object in U; $\mu_N(x, y)$ = strength of the relation between x and y; $\mu_T(x)$ = grade of membership of x in T ([11], p. 136)

Zadeh then regarded a language L as a 'fuzzy correspondence', more explicitly, a fuzzy binary relation, from the term set $T = x$ to the universe of discourse $U = y$ that is characterized by the membership function $\mu_L : T \times U \longrightarrow [0; 1]$. If a term x of T is given, then the membership function $\mu_L(x, y)$ defines a set $M(x)$ in U with the following membership function: $\mu_{M(x)}(y) = \mu_L(x, y)$. Zadeh called the fuzzy set $M(x)$ the meaning of the term x; x is thus the name of $M(x)$.

With this framework Zadeh established in another article the basic aspects of a theory of fuzzy languages ([11], p. 134]). In the following we quote his definitions of fuzzy language, structured fuzzy language and meaning:

Definition: A fuzzy language L is a quadruple $L = (U; T; E; N)$, in which U is a non-fuzzy universe of discourse; T (called the term set) is a fuzzy set of terms which serve as names of fuzzy subsets of U; E (called an embedding set for T) is a collection of symbols and their combinations from which the terms are drawn, i.e., T is a fuzzy subset of E; and N is a fuzzy relation from E (or more specifically, the support of $T(= supp(T) = \{x \mid \mu_A(x) > 0\}$ that is a non-fuzzy subset, to U which will be referred to as a *naming relation* (Fig. 2).

In the case that U and T are infinite large sets, there is no table of membership values for $\mu_T(x)$ and $\mu_N(x, y)$ and therefore the values of these membership functions have to be computed. To this end, universe of discourse U and term set T have to be endowed with a structure and therefore Zadeh defined the concept of a structured fuzzy language.

Definition: A structured fuzzy language L is a quadruple $L = (U; S_T; E; S_N)$, in which U is a universe of discourse; E is an embedding set for term set T, S_T is a set of rules, called syntactic rules of L, which collectively provide an algorithm for computing the membership function, μ_T, of the term set T; and S_N is a set of rules, called the semantic rules of L, which collectively provide an algorithm for computing the membership function, S_N, of the fuzzy naming relation N. The collection of syntactic and semantic rules of L constitute, respectively, the syntax and semantics of L.

To define the concept of meaning, Zadeh characterized the membership function $\mu_N : supp(T) \times U \longrightarrow [0; 1]$ representing the strength of the relation between a term x in T and an object y in U. He clarified:

A language, whether structured or unstructured, will be said to be fuzzy if [term set] T or [naming relation] N or both are fuzzy. Consequently, a non-fuzzy language is one in which both T and N are non-fuzzy. In particular, a non-fuzzy structured language is a language with both non-fuzzy syntax and non-fuzzy semantics ([11], p. 138).

Zadeh now identified these fuzzy subsets of the universe of discourse that correspond to terms in natural languages with its 'meaning':

Definition: The meaning of a term x in T is a fuzzy subset $M(x)$ of U in which the grade of membership of an element y of U is given by $\mu_{M(x)}(y) = \mu_N(x, y)$.

Thus, $M(x)$ is a fuzzy subset of U which is conditioned on x as a parameter and which is a section of N in the sense that its membership function, $\mu_M(x) : U \longrightarrow [0; 1]$, is obtained by assigning a particular value, x, to the first argument in the membership function of N.

Zadeh concluded this paper mentioning that 'the theory of fuzzy languages is in an embryonic stage' ([11], p. 163). In the next sections we will not nurture the field of fuzzy languages but we will reflect and develop some aspects of Fuzziness and Meaning.

1.4

Fuzzy sets' theory basically deals with non-ambiguous statements build up with precise or imprecise predicates P that, through the elemental statements 'x is P', are recognized as being measurable in a crisp universe of discourse X; that is, whose meaning can be described by quantities $(X, \leq_P, m_P)$, where [12]:

1. $\leq_P$ is a binary relation induced in X reflecting the perception-based variation of P in less or in more:

$$x \leq_P y \Leftrightarrow x \text{ is less } P \text{ than } y.$$

   It should be pointed out that in asserting 'x is P', in qualifying x by P, what is actually said is that x shows a property named P; that x carries some information on such property. Hence, $x \leq_P y$ tries to capture that x carries less information on P than carries y. It is thanks to the relation $\leq_P$ that the use of P can appear as being gradable in X.
   Then, P is *measurable* in X if $\leq_P \neq \emptyset$, and P is *meaningless* in X if $\leq_P = \emptyset$. In addition, it can be said that P is *currently metaphysical* in X when seen how such a relation $\leq_P$ can be described [13].
2. A measure m_P is a function $X \longrightarrow [0, 1]$ verifying the three following rules,

   2.1 If $x \leq_P y$, then $m_P(x) \leq m_P(y)$, in the linear order $\leq$ of the unit interval.
   2.2 There is not even If z minimal for $\leq_P$, then $m(z) = 0$.
   2.3 If z maximal for $\leq_P$, then $m(z) = 1$.

The numerical values $m_P(x) \in [0, 1]$ try to reflect the amount of information x carries on P. Like it happens, for instance, with the definition of a Kolmogorov's probability [16] the former three properties don't suffice to specify a single m_P; to specify one of them either more information on the particular use of P in X, or some hypotheses on it, is needed. There is not a unique measure for the extent up to which each x shows the property; meaning is not unique, and P is *effectively measurable* if it is meaningful and at least one of its measures is known.

For instance, the measure of P = big in $X = [0, 10]$, a case in which $\leq_{big}$ is easily identified with the linear order $<$ of this interval [12], it can be either $m_{big}(x) = x/10$, or $m_{big}(x) = x^2/100$, etc. Indeed the general form of this measure is $m_{big}(x) = f(x)$, with a function $f : [0, 10] \longrightarrow [0, 1]$ non-decreasing and verifying $f(0) = 0$ and $f(10) = 1$, that should be specified at each contextual use of P in $[0, 10]$. Notice that, for instance, if taking $f(x) = 0$, if $0 \leq x < 3$, and $f(x) = 1$ otherwise, 'big' could be erroneously identified with 'less than three', a crisp predicate in $[0, 10]$, whose corresponding relation $<_{less3}$ is not coincidental with the order of the interval since 'less than three' is a precise predicate for which, given two numbers x, y in $[0, 10]$, it only can happen that they are, or are not, less than three. Hence, for representing the flexible, linguistic, 'big', function f should be strictly non-decreasing, continuous. If defining $=_P$ as the relation $\leq_P \cap \leq_P^{-1}$, it is $=_{less3}$ an equivalence relation with just the two classes $[0, 3)$ and $[3, 10]$, but $\leq_{big}$ is the linear order of $[0, 10]$. These two predicates are effectively measurable in $[0, 10]$, but their qualitative meanings are different and, consequently, they have different meanings.

As it is known, the nomenclature's change given by:

(I) $x \in_r \mathbf{P} \Leftrightarrow m_P(x) = r$, with $r \in [0, 1]$, read 'x belongs to $\mathbf{P}$ with degree of membership r', jointly with
(II) If P is a precise predicate, then $\mathbf{P}$ is a crisp set, or $m_P(x) \in \{0, 1\}$ for all $x \in X$, and the identity definition,
(III) $\mathbf{P} = \mathbf{Q} \Leftrightarrow m_P = m_Q$,

allow to consider new mathematical objects $\mathbf{P}$, $\mathbf{Q}$, etc., called *fuzzy sets* in X; they specify a meaning of the predicate in the universe of discourse. A fuzzy set $\mathbf{P}$ is called the fuzzy set with *linguistic label* P, and it is obvious that each fuzzy set is completely characterized by its measure, also called its *membership function* [14]. For a given predicate P in X, the multiplicity of fuzzy sets $\mathbf{P}$, its different representations by membership functions m_P, according to the contextual use of P in X, reveals a hidden concept, that of the 'linguistic collective' $\boldsymbol{P}$, of which each $\mathbf{P}$ is but a manifestation. For instance, if P = young is applied in X = London, in the usual linguistic expression 'young Londoners' is hidden the collective $\boldsymbol{P}$ = 'young Londoners', of which each representation by a fuzzy set reflects but a particular state of it. Although the idea of collective is well rooted in language, the fact that predicates 'collectivize' in the universe of discourse still deserves a deep study.

Property 1, does not specify how the values of the measure increase in the unit interval when the elements of X 'grow' in the graph $(X, \leq_P)$; for instance, in the former case of 'big' the measure can increase in linear form $(x/10)$, in quadratic form $(x^2/100)$, etc. In the case of the measures of length, surface, volume, etc., and also in that of probability, this growing is additive. But additivity is just a particular form for such growing corresponding to those cases in which no any loss or gain of information appears when joining two separate pieces of information. In general, measures m_P are not additive, that is, the measure of '(x is P) or (y is P)' is not always the addition of the respective measures of 'x is P' and 'y is P', when both statements are contradictory [15]. Notice that with probability measures applied to elements a, b, c,..., in a Boolean algebra [16], it is $a \leq b'$ (b is contradictory with a) equivalent with $a \cdot b = 0$ (a and b are incompatible) [17], or, with crisp sets A, B,..., it is $A \subseteq B^c \Leftrightarrow A \cap B = \emptyset$, and the *additive law* of probabilities p is stated by: Provided $A \cap B = \emptyset$, it is $p(A \cup B) = p(A) + p(B)$.

The action of P in X is immediately recognized by the given use, or current qualitative meaning, of the elemental, or atomic, statements 'x is P'; it is through the set $\{x \text{ is } P; x \in X\}$ and the relation $\leq_P \subseteq X \times X$, that is possible to capture the primary or qualitative use of P in X; for this reason, the graph $(X, \leq_P)$ is identified with the *primary* or *qualitative* meaning of P in X. At its turn, each value $m_P(x)$ tries to numerically evaluate the amount of information on the property named P carried by the statement 'x is P'. It is for this reason that the *meanings of P in X* are summarized by the quantities $(X, \leq_P, m_P)$, each one corresponding to a full particular use or meaning of P in X. In this way, the abstruse concept of meaning is *ordine geometrico* domesticated, and it appears the possibility of scientifically considering it [18].

Properties 2 and 3 translate what in plain speech is said by 'z is P' is *false*, or *true*, respectively. Actually, the qualified statements 'z is P is false' and 'z is P is true', simply try to indicate that in the corresponding context there is no other x showing P, respectively, less or more than z. In this sense the adjectives 'false' and 'true' can be viewed as scientifically superfluous; in addition, they are charged with a dangerous history when are only 'currently metaphysical'. The metaphysical 'false' and 'true', the mother-predicates of the also metaphysical concept 'Truth', deserve to be seen as a kind of excessively ambiguous pseudo-concept that, probably, cannot be easily expulsed from ordinary speech.

Zadeh's fuzzy sets [14] try to add to the theory of classical sets, allowing the specification of precise terms, a way for mathematically representing imprecise linguistic, but not ambiguous, terms. The important linguistic phenomenon of ambiguity is not yet scientifically domesticated; if an imprecise term P can be domesticated from empirically and perceptively capturing a single qualitative meaning exhibited by a graph $(X, \leq_P)$, domesticating ambiguous terms Q, those that show several meanings, will require, perhaps, to start with more than one graph. It is an open problem.

It should be noticed that if just working with the measures m, as it is usual in the applications of fuzzy sets, then instead of the relation $\leq_P$ only the relation defined by

$$x \leq_m y \Leftrightarrow m(x) \leq m(y),$$

can be taken into account. Since this new relation, called the *working meaning* of the predicate, is a linear one ($\leq$ is indeed a total order in the unit interval), but $\leq_P$ is not always linear, both relations cannot always coincide. From the obvious implication: $x \leq_P y \Rightarrow x \leq_m y$, it results $\leq_P \subseteq \leq_m$: usually the working meaning is larger than the primary meaning [12]. Provided $\leq_P$ is not a total, or linear, relation, the act of introducing a measure m_P enlarges the qualitative meaning expressed by $\leq_P$. The act of effectively measuring a meaningful predicate enlarges its meaning. The why for this is clear: to specify a particular measure, and further of knowing the qualitative use of the predicate, information on its graded variation is necessarily required.

1.5

In classical two-valued propositional logic, the concept of negation (also called the 'logical complement') that operates by taking a proposition p to another proposition 'not p' (we will note $\neg p$).

In predicate logic that is an extension of propositional logic by using quantors and developed by Frege in his *Begriffsschrift* (1879) [19] and independent by Charles S. Peirce (see [20]). In predicate logic two quantifiers are defined: the universal quantifier, $\forall$, and the existential quantifier, $\exists$.

> $\forall$ is interpreted as 'for all'. It expresses that a proposition can be satisfied by every member of a domain of discourse.
> $\exists$ is interpreted as 'there exists'. It expresses that a propositional function can be satisfied by at least one member of a domain of discourse.

With these quantifications also the concept of opposition is much more interesting because now we have to distinguish between three different logical relations of opposition and negation is linked to the relation of contradiction which is one of them:

(1) A *contradictory* statement of a statement has the opposite truth value, i.e. the two statements cannot be true together and they cannot be false together.
(2) A *contrary* statement of a statement is a statement that cannot be true if the original statement is true. However, we notice that they can both be false.
(3) A *subcontrary* statement of a statement is a statement that cannot be false if the original statement is false.

Usually, these opposition-relations are shown in the traditional square of opposition. What is said in this figure we can find already in the writings of Aristotle (*De interpretatione*) (however he never showed such a figure) and also Frege used it (see Fig. 3):

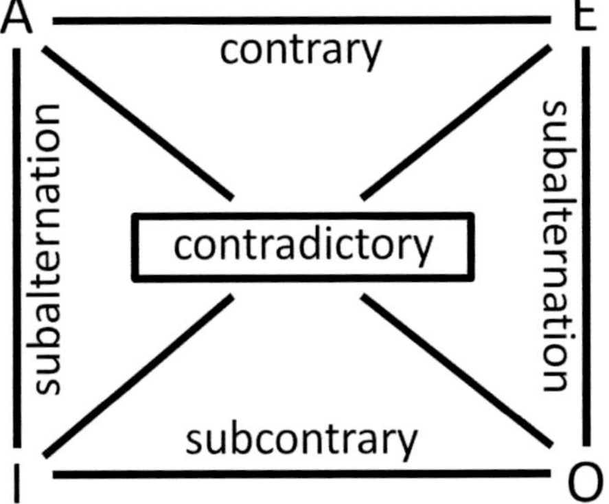

Fig. 3 Logical square of opposition

We explain what is represented in that figure: Letters A, E, I, and O represent the four classic forms of propositions (or statements) of predicate logic:

A: a general affirmative statement of the form: All S are P.
E: a general negative statement of the form: No S are P.
I: a particular affirmative statement of the form: Some S are P.
O: a particular negative statement of the form: Some S are not P.

Given the assumption made within classical (Aristotelian) categorical logic, that every category contains at least one member, the following relationships, depicted on the square, hold.

The square of opposition shows:

(1) Statements A and O are contradictory, and also statements E and I are contradictory.
(2) Statements A and E are contrary.
(3) Statements I and O are subcontrary, and also statements I and O are subcontrary, but they are not contrary or contradictory.
(4) The relation of subalternation says that the truth of the first statement implies the truth of the second statement. Statement A stands in the subalternation relation with the corresponding I statement and statement E is in that relation with statement O.

Furthermore, among these logical oppositions there is another opposition that we have to consider from the linguistic point of view. This kind of opposition is what is called antonym and with it also the concept of meaning comes into the play again: An antonym is defined to be a word having a meaning opposite to that of another word.[5]

Following linguistic research [21] we identify three different types of antonyms:

- *Gradable antonyms.*
  These antonyms are word pairs whose meanings are opposite and which lie on a continuum, e.g.: (very) big, (very) small; hot, cold.
- *Complementary antonyms.*
  These antonyms are word pairs whose meanings are opposite but whose meanings do not lie on a continuous spectrum; they express an either/or relationship, e.g.: dead or alive, male or female; push or pull.
- *Converse or relational antonyms.*
  These antonyms are word pairs where opposite makes sense only in the context of the relationship between the two meanings. They express reciprocity, e.g. borrow or lend, buy or sell, wife or husband, teacher or pupil.

In the following subsections we only consider the first kind of antonyms: gradable antonyms.

---

[5]See for instance: Webster dictionary, URL: http://www.merriam-webster.com/dictionary/antonym.

1.6

In the learning of a natural language it is difficult, if not impossible, to capture the meaning of 'x is P' without simultaneously capturing those of 'x is $n(P)$' and 'x is $a(P)$', with $n(P)$ the negation of P, and $a(P)$ an antonym or opposite of P [22]. For instance, how to recognize that 'John is tall', without being able to recognize that 'Peter is short'? This is the reason for which it is important to know if being P measurable, so are $n(P)$ and $a(P)$. Language shows these two ways, the second weaker and more imprecise than the first, for separating what lies under P from what is not under P in a given universe of discourse X. Language makes two different cuts between 'what is under P' and 'what is not under P', the first by the particle 'not', and the second by an antonym; think, for instance, with P = old, in $a(P)$ = young and $n(P)$ = not old, where with old and young is created the mixed and common term 'middle aged' as 'neither old, nor young', whereas 'old and not old' seems to express a more strong separation. Language always tends to find a flexible use of its terms, but logic is much more strict and inflexible; the use of words exhibits a superposition of meanings with their antonyms and negation similar, up to some extent, to the quantum phenomenon of state's superposition.

If P is a linguistic term, in short, it is in the dictionary, also $a(P)$ is so, but $n(P)$ is not. For instance both 'tall' and 'short' are in the dictionary, but neither 'not tall', nor 'not short' are in it. Hence it has no sense to search for '$a(n(P))$', except if admitting that it coincides with '$n(a(P))$'; that is, accepting the 'commutation' $a \circ n := n \circ a$, of the operators a and n. For instance, if P = full, a (not full) = $n(a(P))$ = n (empty) = not empty := $a(n(P))$. This is a subject deserving further analysis in language, similar to that needed to know if to negate a statement like 'Most rich people are avaricious', in plain language it is either preferred 'Few rich people are avaricious', or 'Most rich people are not avaricious'.

As it can be checked with the same predicate 'full', it is always 'If x is $a(P)$, then x is $n(P)$', but not reciprocally; symbolically, $a(P) \Rightarrow n(P)$, that is, the negation can be seen as a semantic and inaccessible upper limit of the opposites [22]. It is the same with 'If it is short, then it is not tall', but it is not that 'If it is not tall, it is short', short implies not tall, but not tall does not imply short. Only when the language did not previously require to consider any opposite of P, as it happens with some technical terms like, for instance, 'normally distributed', is when it can be irregularly taken $a(P)$ as artificially coincidental with $n(P)$. For instance, there are dictionaries of antonyms where the antonym of the term 'probable' appears as 'not probable', instead of the 'improbable' appearing in other dictionaries that, nevertheless, just describe it as 'not probable' (Webster's dictionary, for instance).

Concerning the measures of $a(P)$ and $n(P)$, provided they exist, the inequality $m_{a(P)} \leq m_{n(P)}$, can be stated and seen as a coherence condition [22]. Since, and contrarily to the non-uniqueness of the antonyms, there is just one linguistic negation n (P), its measure can be seen as an upper limit of those of the antonyms. The opposite affirms, but the negation denies and often deserves the question 'why?' typical of children when receiving a 'not' command, but that is not always asked when receiving a command with an antonym. If such upper limit exists and is different from $m_{P'}$, it can be seen as a 'linguistic complement' of P; an open problem that, if having some algebra, it is yet unknown as it is its own existence.

Note

Fuzzy sets algebras are constructed over the set $[0, 1]^X$, like the (naïve) theory of classical sets is built on $\{0, 1\}^X$ with the structure of Boolean algebra isomorphic to that of the power-set $\mathcal{P}(X)$ when endowed with intersection, union, and complement. It should be pointed out that any algebra of fuzzy sets, like, for instance the general concept of a Basic Fuzzy Algebra [23], should preserve, when restricted to $\{0, 1\}^X$, the Boolean algebra of crisp sets. In the case of **n (P)** there is no problem, since if **P** is a crisp set, that is, if $m_P(x)$ is in $\{0, 1\}$ for all x in X, then also the values $m_{P'}(x)$ belong to $\{0, 1\}$ and reproduce the classical complement $\mathbf{P}^c$ of **P**. Nevertheless, with **a (P)** a problem arises since crisp algebras do not consider the antonym's operation. This is another subject deserving further analysis; for instance, in [0, 1], 'one half' represented by the singleton $\{0.5\}$, has the negation 'not one half', represented by the union $[0, 0.5) \cup 0.5, 1]$, but the antonym of 'one half' is unknown in mathematical language.

The algebras in $[0, 1]^X$ are constructed with an 'intersection', a 'union' and a 'complement' of fuzzy sets, trying to translate into fuzzy sets the connectives 'and', 'or' and 'not' between elemental statements in such a form able to represent them in many contexts; for this purpose the algebra should be submitted to a minimal number of laws [23] in comparison to the strongest Boolean algebra's structure of crisp sets, and adding more laws at each particular situation. The fact is that there is not a unique algebra of fuzzy sets, like it is only one of crisp sets. At each case, an algebra adapted to the corresponding context should be designed, and what is not yet known is how to incorporate, and with which laws, the antonyms. There is not an specification axiom like that of naïve set theory [24]; to specify a fuzzy set for a given linguistic label is more complicated since usually it also requires not only to specify an antonym of it, but also a particular algebra of the fuzzy sets representing the other involved linguistic labels. There is not an exact axiom of specification analogous to that of set theory that assures the existence of a single crisp subset for each binary (say, non gradable) predicate on a universe of discourse X; crisp sets can be seen as rigid entities, but fuzzy sets as flexible ones. In the precise, rigid, case, the collective has just a single state, but in the imprecise, flexible, it has many.

2 Measurability of $a(P)$ and $n(P)$

Provided P is effectively measurable by a quantity $(X, \leq_P, m_P)$, what can be said on the measurability of $a(P)$ and $n(P)$? From now on let us shorten $a(P)$ by aP, and $n(P)$ by P'.

2.1

The idea of term's opposition P/aP refers to the existence of a symmetry in the universe of discourse, $s : X \to X$, such that: 'x is P' if and only if '$s(x)$ is aP'; $s^2 = s$; $x \leq_P y \Leftrightarrow s(y) \leq_P s(x)$, and $\leq_{aP} = \leq_P^{-1}$, that is [12],

$$x \leq_{aP} y \Leftrightarrow y \leq_P x, \text{ or } \leq_{aP} = \leq_P^{-1} .$$

If z is a maximal (minimal) for $\leq_{aP}$ then $s(z)$ is a maximal (minimal) for $\leq_P$. Thus, not only the graph $(X, \leq_P^{-1})$ reflects the primary meaning of aP in X, but the mapping $m_{aP} = m_P \circ s$, verifies:

1. $x \leq_{aP} y \Leftrightarrow y \leq_P x \Rightarrow s(x) \leq_P s(y) \Rightarrow m(s(x)) \leq m(s(y)) \Leftrightarrow (m_P \circ s)(x) \leq (m_P \circ s)(y)$,
2. If z is a minimal for $\leq_{aP}$: $(m_P \circ s)(z) = m_P(s((z)) = 0$, since $s(z)$ is also a minimal for $\leq_P$,
3. If z is a maximal for $\leq_{aP}$: $(m_P \circ s)(z) = m_P(s(z)) = 1$, since $s(z)$ is a also a maximal for $\leq_P$,

and the quantity $(X, \leq_P^{-1}, m_P \circ s)$ represents a meaning of aP in X directly following from that of P. Each antonym requires its own symmetry.

It should be noticed that it has only been proven that the graph $(X, \leq_P^{-1})$ expresses the qualitative meaning of aP in X, and that there exists, at least, a kind of measures depending on the symmetries s, actually giving measures for aP once a measure for P is known. What it has not been proven at all is that any quantity reflecting a meaning for aP should belong to this type; this remains an open problem.

To conclude, if P is measurable also aP is so, and particular measures can be found for aP. Hence, from the effective measurability of P follows that of aP.

2.2

The idea of the negation P' of P is more sophisticated than that of an opposite since the only that can be surely asserted is that [12],

$$x \leq_P y \Rightarrow y \leq_{P'} x, \text{ but not reciprocally,}$$

that is: $\leq_P^{-1} \subseteq \leq_{P'}$, but what not always can be asserted is $\leq_{P'} \subseteq \leq_P^{-1}$. From this inclusion what just follows is that $\leq_{P'}$ is not empty, that is, provided P is measurable, P' is neither metaphysical, nor meaningless, but measurable. Of course, what is not immediately clear is the relationship between minimals and maximals between both relations $\leq_P$ and $\leq_{P'}$ since, in general, the second has more 'arcs' that the first. It is not always immediately sure that what is *true* for P should be *false* for P', and what is *false* for P *true* for P'.

Anyway, what can be stated, but only in some cases, is the existence of mappings $m_{P'}$, from X into the unit interval, verifying the three properties of a measure; for instance and provided $\leq_{P'}$ coincides with $\leq_P^{-1}$, the definition $m_{P'} = 1 - m_P$, verifies:

$$x \leq_{P'} y \Leftrightarrow y \leq_P x \Rightarrow m_P(y) \leq m_P(x) \Leftrightarrow$$
$$\Leftrightarrow 1 - m_P(x) \leq 1 - m_P(y) \Leftrightarrow m_{P'}(x) \leq m_{P'}(y),$$

and keeps the 0 and 1 values for minimals and maximals. Thus, in this case the quantity $(X, \leq_P^{-1}, 1 - m_P)$ reflects a meaning of P', P' is effectively measurable. Notwithstanding, it seems that nothing can be asserted for the three properties when the two relations do not coincide, in which case the effective measurability of P' is still an open problem.

Being not the negation a linguistic term, it cannot be surprising at all to find difficulties in founding measures for P' in all cases; for instance, it is not clear when the negation should depend on the particular x that is considered. Notice that when P'' coincides with P, from $\leq_{P'}^{-1} \subseteq \leq_{P''} = \leq_P$, it just follows the already accepted inclusion $\leq_P^{-1} \subseteq \leq_P$.

Since aP is always measurable, each time in which also $n(aP)$ is so, the mapping $1 - m_P \circ s$, is a measure for it. Notice that, instead of the function $1 - id$, it can be also taken any order-reversing numerical function $N : [0, 1] \to [0, 1]$ such that $N(0) = 1$ and $N(1) = 0$, giving the measure $N \circ m_P$ for P'. Such functions are called *negation functions* [25, 26]. Notice also that provided it is accepted that n and a commute, then the measures of $a(nP))$ and $n(aP))$ do coincide; for instance the measure of a(not young) will coincide with that of not(old); nevertheless, this deserves a further study anchored in language.

2.3

If it seems clear, from the linguistic experience, that it is always $aaP = P$ (for instance, *aa* empty = *a* full = empty), that is, that $a^2 = id$; nevertheless, it is not so clear if it can be always $P'' = nnP = P$, that is $n^2 = id$. There are well known cases in which the only that can be accepted is that 'If x is P, then x is P'''. Those negations for which it is always $P'' = P$, are called *strong negations*; in these cases the former functions N obviously should verify $N^2 = id$.

For what concerns the law $a^2 = id$, notice that when symmetries s_P and s_{aP} exists giving $m_{aP} = m_P \circ s_P$, and $m_{aaP} = m_{aP} \circ s_{aP}$, it follows $m_{aP} = (m_P \circ s_P) \circ s_{aP} = m_P \circ (s_P \circ s_{aP})$, and it suffices to have $s_P = s_{aP}$. If a symmetry s gives a measure of aP, then the same symmetry is giving that of aaP.

In the case of P', such that $P'' = P$, and if there are two functions N_P and $N_{P'}$, such that $m_{P'} = N_P \circ m_P$, and $m_{P''} = N_{P'} \circ m_{P'}$, it follows $m_{P''} = N_{P'} \circ (N_P \circ m_P) = (N_{P'} \circ N_P) \circ m_P$, implying $N_{P'} = N_{P''}$. If a strong negation gives a measure of P', then the same strong negation is giving that of P''.

When P' is effectively measurable with a measure $m_{P'}$, the above condition of coherence is $m_{aP} \leq m_{P'}$. If these measures are given by a symmetry s_P, and a negation function N_P, such inequality is

$$m_P \circ s_P \leq N_P \circ m_P \Leftrightarrow m_P \leq N_P \circ m_P \circ s_P,$$

showing that the symmetry in X for aP, and the negation function in $[0, 1]$ for P', cannot be independently chosen the one from the other [22]. If $s_P(N_P)$ is known, $N_P(s_P)$ must satisfy the former inequality.

2.4

Often language introduces the *middle term* **m** $P =$ '$n(p)$ and $n(aP)$'. For instance, 'warm' means 'not-hot and not-cold', with cold = a (hot). Provided P' is effectively measurable, 'and' is represented [23] by a continuous t-norm T, 'not' by a strong negation N, and the antonym by a symmetry s, the measure of the middle term **m** (P) is given by $m_{mP} = T \circ (mP' \times N(m_{aP})) = T \circ ((N \circ m_P) \times (Nm_P \circ s))$.

At each case, the functions T, N, and s should be taken accordingly with the corresponding context, and the measure m_P accordingly designed [22].

2.5

If from the measurable character in X of P follows that of P', and since it always follows that of any antonym aP of P, the fuzzy sets **P**, **P'** and **a P**, respectively defined by the membership functions m_P, mP' and m_{aP}, can be considered when they exist; namely, when P' is effectively measurable.

In fuzzy set theory it is usually defined that a fuzzy set **A** in X is *contained* in another **B** also in X, $\mathbf{A} \subseteq \mathbf{B}$, if and only if $m_A \leq m_B$. Notice that from this definition follows $\mathbf{A} = \mathbf{B} \Leftrightarrow \mathbf{A} \subseteq \mathbf{B}$ & $\mathbf{B} \subseteq \mathbf{A}$. Hence, it follows that $\mathbf{a\,P} \subseteq \mathbf{P'}$, from the coherence condition $m_{aP} \leq m_{P'}$, that allows to see each **a P** as a fuzzy subset of the 'complement' **P'**, provided P' is effectively measurable.

From the coherence condition it follows the existence of the number,

$$\text{Sup}\ \left\{m_{aP}(x)\ ;\ \text{for all } aP \text{ and all } x \in X\right\} = m_{c(P)}(x) \leq m_{P'}(x),$$

and the fuzzy set **CP**, defined, if it exists, by the membership function $m_{c(P)}$ can be seen as a linguistic kernel of the complement that can have or have not a linguistic term label. Such kernel can or cannot coincide with **P'** and it appears the interesting problems of studying when they can coincide, the differences between **CP** and **P'** if they are not coincidental, and what happens if changing the theory of what is known with the 'classical' complement **P'** by the new fuzzy set **CP**. Another, but still unexplored way, for arriving at a kind of core, or kernel, of what opposes P and is comprised by **P'** could consist, when there are a few number of known opposites as it is usually the case, in defining a kind 'envelope' of these opposites included in the fuzzy set labelled P'. Notwithstanding, these are new open problems, as it is that of knowing a suitable linguistic label CP for **CP**, and the existence of not of a relation $\leq_{CP}$ for which the function $m_{C(P)}$ can be a measure.

3 Strong Negation Functions

3.1

Fuzzy logic mostly supposes that the negation always verifies $P'' = P$, avoids the problem around $\leq_{P'}$, presupposes the existence of a measure $m_{P'}$ that considers is functionally expressible in the form $N \circ m_P$, with N a fixed strong negation function, and just takes into account the working meaning given by $m_{P'} = N \circ m_P$. To some extent and currently in praxis, fuzzy logic avoids to do a deep semantic analysis of the used negation that, if not actually important for many applications, can be of a dubious effect when, in the way of Zadeh's *Computing with Words*, more complex and larger pieces of words will need to be represented through capturing its contextual and purpose-driven meaning.

One of the advantages of strong negations is their continuity; because of the property $N^2 = id$, or $N = N^{-1}$, implies that these functions are strictly non-decreasing and continuous. Hence, they will never introduce more discontinuities in the membership function of P' than those that were already in the membership function of P. One of the disadvantages of fixed strong negations is the usually not explained hypothesis that $m_{P'}$ should be functionally expressible from m_P. This lack of explanation also comes not only from a lack of analysis of the semantics of the negation, and from the convenience of not adding new discontinuities, but from the easiness in managing strong negation functions. This easiness comes from the functional-parametric characterization of these functions N given by the theorem [25]:

- A function $N : [0, 1] \to [0, 1]$ is order-reversing, involutive, and verifies $N(0) = 1$, if and only if it exists an order-auto-morphism φ of the totally ordered unit interval, such that $N = N_\varphi = \varphi^{-1} \circ (1 - id) o\varphi$, where function φ is the functional parameter.

Thus, there are no other strong negation functions than those N_φ belonging to the family of the 'basic' strong negation $N_{id}(x) = 1 - x$. For instance,

- If $\varphi(x) = x^p$, with p a real number, it follows $N_p(x) = (1 - x^p)^{1/p}$, of which the case $p = 2$ gives the 'circular' negation $N_2(x) = \sqrt{(1 - x^2)}$.
- If $\varphi(x) = log(1 + x^p)^{1/q}$, with $q > -1$ and $p > 0$, follows the bi-parametric family of strong negations $N_{p,q}(x) = (1 - x^p/1 + qx^p)^{1/p}$, that contains the only family of rational strong negation functions given by $p = 1 : N_q(x) = 1 - x/1 + qx$, with $q > -1$, and called the Sugeno's family of strong negations. With $q = 0$ is obtained the 'basic' strong negation $N_0(x) = 1 - x$, that is the only strong negation that is a linear function.
- It should be noticed that $N_\varphi = N_\delta$ does not imply $\varphi = \delta$. In general, each strong negation function can be expressed by a family of order-auto-morphisms φ, δ, etc., of the ordered unit interval.

Notice that an auto-morphism φ is nothing else than a strictly non-decreasing function $[0, 1] \to [0, 1]$, such that $\varphi(0) = 1 - \varphi(1) = 0$, and of which the functions $\varphi(x) = x^n$, n a natural number, are typical examples.

3.2

The equation $N(x) = x$ has a unique solution $x = n$, there is a single fix-point n of N [23], since N is continuous; this point n is in the open interval $(0, 1)$ since it is $N(0) = 1$ and $N(1) = 0$. If $N = N_\varphi$, It is $n = \varphi^{-1}(1/2)$:

$$\varphi^{-1}(1 - (x)) = x \Leftrightarrow 1 - \varphi(x) = \varphi(x) \Leftrightarrow \varphi(x) = 1/2.$$

For instance, in the case of the Sugeno family with $q > 0$, the fix-point is $n_q = (\sqrt{(1 + q)}) - 1)/q$. Obviously, it is

$$x \leq n \Leftrightarrow n \leq N(x), \text{ and } n \leq x \Leftrightarrow x \leq N(x).$$

A fuzzy set **P** is self-contradictory provided it is **P** $\subseteq$ **P'**, that is, provided P' is effectively measurable and **P'** functionally expressible by a strong negation function. It holds

$$m_P \leq N_\varphi \circ m_P \Leftrightarrow \varphi \circ m_P \leq 1 - \varphi \circ m_P \Leftrightarrow m_P(x) \leq_\varphi^{-1} (1/2),$$

for all x in X. Self-contradictory fuzzy sets are those whose membership function values are not over the fix-point of the strong negation function. For instance, if **P** is a crisp set and since in this case it is $m_P(x) \in \{0, 1\}$ for all x in X, it is

$$m_P(x) \leq \varphi^{-1}(1/2) \in (0, 1), \text{ for all } x \text{ in } X \Leftrightarrow m_P(x) = 0,$$

that is $\mathbf{P} = \emptyset$: the only self-contradictory crisp set is the empty one, and perhaps for this reason is not immediate to accept the empty set as a true set.

There is a multiplicity of strong negation functions with the same fix-point in $(0, 1)$ [26, 27]. For instance, the following family of strong negations is made of piecewise linear functions N_r with the single parameter $r \in (0, 1)$,

$N_r(x) = ((r-1)/r)x + 1$, if $x \in [0, r]$;
$N_r(x) = r(x-1)/(r-1)$, if $x \in [r, 1]$,

with which $N_{0,5}(x) = 1 - x$, and for all r is $N_r(r) = r$.

Since all points in the unit open square $(0, 1) \times (0, 1)$ can be a fix-point for a multiplicity of negations, it is almost obvious that it does not exist neither a minimum negation, nor a maximum one in the pointwise ordering of strong negations given by $N^1 \leq N^2 \Leftrightarrow N^1(x) \leq N^2(x)$, for all x in X.

There are, of course, negation functions that are continuous and strictly non-decreasing, but not strong, for instance, $N^*(x) = 1 - x^2$, for which it is $N^*(N^*(x)) = x^2 \leq x$. These functions, are called *strict* [26] and have a doubtful theoretical interest within the theory of fuzzy sets since they lack of a known characterization like it happens with the strong ones.

If δ is an order auto-morphism of the unit interval, and N is a strict negation, also $N^\delta = \delta^{-1} \circ N \circ \delta$, is a strict negation that is strong if and only if N is strong [26]. In this case, if N is given by an auto-morphism φ, $N = N_\varphi$, from $N^\delta = \delta^{-1} \circ (\varphi^{-1} \circ (1 - id) \circ \varphi)) \circ \delta = (\delta^{-1} \circ \varphi^{-1}) \circ (1 - id) \circ (\delta \circ \varphi)$, and $\delta^{-1} \circ \varphi^{-1} = (\varphi \circ \delta)^{-1}$, follows $N^\delta = N_{\varphi \circ \delta}$. There is easy to obtain strong negations through order auto-morphisms; or instance, if $N(x) = 1 - x/1 + x$, and $\delta(x) = x^3$, it is $N^\delta(x) = (1 - x^3/1 + x^3)^{1/3}$, and if $N(x) = 1 - x$ and $\delta(x) = 2x/1 + x$, it is $N^\delta(x) = 1 - x/1 + 3x$, the negation N_3 among those in the Sugeno's family although given by a different auto-morphism that the one, $\varphi(x) = log(1 + x)^{1/3}$, with which these strong negations were defined. This is, at its turn, an example showing that the automorphism defining a strong negation is not unique.

3.3

The most general form of obtaining a non-functionally expressible strong negation [27] seems to be by means of an expression $m'(x) = N_{m,x}(m(x))$, with a family of functions $N_{m,x} : [0, 1] \to [0, 1]$, depending on both the membership function m in $[0, 1]^X$, and the point x in X. Nevertheless, not all these functions $N_{m,x}$ can be strong negations; it is not too difficult to prove that some of they cannot be so [27].

Anyway, given a family of strong negations N_x, $x \in X$, just depending on the points x in X, but not on m, the transform given by

$$m'(x) = N_x(m(x)), \text{ for all membership functions } m \in [0, 1]^X,$$

gives a strong negation:

(a) $m_1 \leq m_2 \Leftrightarrow m_1(x) \leq m_2(x)$, for all x in $X \Rightarrow N_X(m_2(x)) \leq N_x(m_1(x)) \Leftrightarrow m'_2 \leq m'_1$,
(b) If $m(x) = 0$, $m'(x) = N_x(0) = 1$; if $m(x) = 1$, $m'(x) = N_x(1) = 0$,
(c) $m''(x) = N_x(m'(x)) = N_x(N_x(m(x))) = m(x) : m'' = m$.

Hence, it seems that it is only through families of strong negations N_x as non-functionally expressible strong negation functions can be reached. It should be noticed that the fix-point of a strong negation N_x is not a point but a function $(0, 1) \to (0, 1)$, of $x : N_x(n) = n \Leftrightarrow N_{\varphi x}(n) = n \Leftrightarrow n = \varphi_x^{-1}(x)$. For instance, the point-dependent strong negation defined by,

$N_x(x) = 1 - x$, if $x \in [0, 0.5)$,
$N_x(x) = 1 - x/1 + x$, if $x \in [0.5, 1]$,

has the family of fix-points following from the equation $N_x(n_x) = n_x$, and shown by the graphic in the unit square with the two linear pieces:

$n_x = 0.5$, if $x \in [0, 0.5)$,
$n_x = \sqrt{2 - 1}$, if $x \in [0.5, 1]$.

Of course, the fuzzy sets in $[0, 1]^{[0,1]}$, with membership function m that are self-contradictory are, in this case, those verifying $m(x) \leq n_x$ for all point x in $[0, 1]$. That is, the membership functions that are below 0.5 in $[0, 0.5)$, and below $\sqrt{2 - 1} \sim 0.4142$ in $[0.5, 1]$.

3.4

By mixing strong negations and symmetries, that is, by defining [28] Ovchinnikov's general negations $m_P^*(x) = N_X(m_P(s(x))$, with $N_X = N_{S(X)}$, for all x in X, there is the danger of non keeping the classical negation when m is in $\{0, 1\}^X$, that is, when m represents a crisp set in X.

For example, the definition $m^*(x) = 1 - m(1 - x)$, for m in $[0, 1]^{[0,1]}$, in which it is $s = N^X = 1 - id = N_{s(X)}$ for all x in $[0, 1]$, gives a mapping $^* : [0, 1]^{[0,1]} \to [0, 1]^{[0,1]}$, that verifies all the properties of a strong negation and that if m is crisp also m^* is crisp, but that this m^* not always coincides with the classical complement of m. For instance, if m is the membership function of the crisp set $A = [0.4, 0.5]$, it is m^* the membership function of the crisp set $A^* = [0, 0.5) \cup (0.6, 1]$, that is not the classical complement $A^c = [0.4, 0.5]^c = [0, 0.4) \cup (0.5, 1]$ but contains it.

Notice that what the above membership functions $m_P^* = N_x \circ m_P \circ s$ represent is the membership of *n*(*a*(*P*)), but not that of *n*(*P*). Hence it is not at all surprising that some troubles concerning the preservation of the classical case can appear if m_P^* is confused with $m_{P'}$.

The possibility of 'losing the crisp case' is a trouble paper [28], a mathematically excellent one for another side, shows for the algebra of fuzzy sets in mixing negation and antonym.

3.5

When the intersection and the union of fuzzy sets are expressed by families of continuous t-norms and t-conorms depending on the point, and the strong negation is just given by families of point-dependent strong negations, the known results on the principles of non-contradiction and excluded-middle in Standard algebras of fuzzy sets, can be immediately generalized [27]. That is, if working in the Basic Fuzzy Algebra $([0, 1]^X, \cdot, +, ')$ given by:

$(A \cdot B)(x) = T_x(A(x), B(x))$,
$(A + B)(x) = S_x(A(x), B(x))$, and
$A'(x) = N_x(A(x))$, for all x in X,

then,

Theorem 1 *It is* $(A \cdot A')(x) = 0$ *for all x in X, if and only if*

(1) All t-norms T_x *are in the family of continuous t-norms of the form* $\varphi^{-1} \circ W \circ (\varphi \times \varphi)$*, where W is the Łukasiewicz t-norm* $W(a, b) = max(0, a + b - 1)$*, and* φ *is an order auto-morphism of the ordered unit interval,*
(2) $N_x \le N_{\varphi x}$*, provided* $T_x = \varphi_x^{-1} vW \circ (\varphi_x \times \varphi_x)$.

Notice that when all t-norms are a fixed one *T*, and all negations are a fixed one *N*, this theorem gives the well known one in the Standard algebras of fuzzy sets.

Theorem 2 *It is* $(A + A')(x) = 1$ *for all x in X, if and only if*

(1) All t-conorms S_x *are in the family of continuous t-conorms of the form* $\pi^{-1} \circ W^* \circ (\pi \times \pi)$*, where* W^* *is the t-conorm* $W^*(a, b) = min(1, a + b)$*, and* π *is an order automorphism of the ordered unit interval,*
(2) $N_{\pi x} \le N_x$*, provided* $S_x = \pi_x \circ W^* \circ (\pi_x \times \pi_x)$.

Analogously, when all t-conorms are a fixed one *S*, and all negations are a fixed one *N*, this theorem gives the well known one in the Standard algebras of fuzzy sets [29].

Corollary 1 *The only algebras* (T_x, S_x, N_x) *in which the two principles of non-contradiction and excluded-middle jointly hold, are those with all the t-norms and t-conorms of the forms* $T_x = \varphi_x \circ W \circ (\varphi_x \times \varphi_x)$*, and* $S_x = \pi{-}1_x \circ W^* \circ (\pi_x \times \pi_x)$*, and verifying the nesting inequality* $N_{\pi x} \le N_x \le N_{\varphi x}$*, for all x in X.*

Of course, whenever both families of auto-morphisms are coincidental, that is, when $\varphi_x = \pi_x$, for all x in X, the last condition is reduced to: $N_x = N_{\varphi x}$.

Note

Some logically driven theoreticians [30], derive the negation for fuzzy sets from the so-called residuated implication functions

$$J_T(a,b) = Sup\,\{z[0,1]; T(a,z) \leq b\}\,,$$

where T is a continuous t-norm, and by defining $N_T(a) = J_T(a,0)$.

These operators come directly from the Boolean identity $a' + b = Sup\,\{z \in B; a \cdot z \leq b\}$, only valid in complete Boolean algebras $(B, \cdot, +, ')$ as a consequence that implication operations $\rightarrow$ in B, defined by verifying the so-called Modus Ponens inequality $a \cdot (a \rightarrow b) \leq b$, that holds if and only if $a \rightarrow b \leq a' + b$. Of course, in Boolean algebras it is $a' = a' + 0 = a \rightarrow b$, as well as in Orthomodular lattices with the Sasaki hook $a \rightarrow b = a' + a \cdot b$, or the Dishkant hook $a \rightarrow b = b + a' \cdot b'$, is also $a' = a \rightarrow 0$.

- If $T = min$, it is $J_{min}(a.b) = 1$, if $a \leq b$; and $J_{min}(a,b) = b$, if $a > b$. Hence, $N_{min}(a) = J_{min}(a,0) = 1$, if $a = 0$, and 0 if $a > 0$, that is a discontinuous strong negation.
- If T is in the family of product, $T = \varphi^{-1} \circ Prodo(\varphi \times \varphi)$, is $J_T(a,b) = 1$, if $a \leq b$; and $J_T(a,b) = \varphi^{-1}(\varphi(b)/\varphi(a))$, if $a > b$. Hence, it is also $N_T(a) = J_T(a,0) = N_{min}(a)$.
- If T is in the family of Łukasiewicz, $T = \varphi^{-1} \circ W \circ (\varphi \times \varphi)$, with the t-norm $W(a,b) = max(0, a + b - 1)$, is $J_T(a,b) = \varphi^{-1}(min(1, 1 - \varphi(a) + \varphi(b))$, and $N_T(a) = J_T(a,0) = \varphi^{-1}(1 - \varphi(a)) = N\varphi(a)$.

All that shows that under this interpretation, that comes from classical Boolean logic considering the material implication given by 'not a or b' $(a' + b)$, strong negation functions are linked with a very particular form of representing the rules 'If x is P, then y is Q'. It should be recalled that the residuated implications [31] with the t-norm in the Łukasiewicz family, are the only that are simultaneously residuated and S-implications $J(a,b) = S(1 - a, b)$, with a t-conorm in the family of $W^*(a,b) = min(1, 1 - a + b)$. Of course, if representing the rules by an S-implication, it is $N(a) = S(1 - a, 0) = 1 - a$; strong negations just appear in this limited view as coming from t-norms in the Łukasiewicz family.

All that is but a typical reminiscence of classical logic. Notice, that if the rules (conditional statements) were represented by a Mamdani-Larsen conditional, $J(a,b) = T(a,b)$ with $T = min$, or $T = Prod$, as it is usual in fuzzy control, it is $J(a,0) = 0$, and no negation is obtained. The interests of logic are mainly placed in the field of formalized artificial languages, but are not in that of the natural one in which there is no a universal representation of conditional statements in which, in most cases, is not well known what is the negation of its antecedent.

It should be finally noticed that the negation N_{min} is strictly smaller than all strong negations N_φ, since $N_{min}(0) = 1 = N_\varphi(0)$, and if $a > 0$, $N_{min}(a) = 0 \leq N_\varphi(a)$; that is, $N_{min} \leq N_\varphi$, and, since N_{min} is discontinuous, it cannot coincide with N_φ. Hence, $N_{min} < N_\varphi$.

4 Conclusions

4.1

This paper, devoted to reflect around what is 'not *P*' and what is 'opposite to *P*', comes from old worries of one of its authors inspired on a chain of three old papers, published in the late seventies and early eighties of last XX century, that can be considered as pioneers in the subject of strong negations in fuzzy set algebras. In the first, Robert Lowen posed in 1978 [32] the problem of obtaining the negation m' of a fuzzy set m by means of numerical functions $N : [0, 1] \to [0, 1]$ verifying $N(0) = 1, N(1) = 0$, if $a \leq b$, then $N(b) \leq N(a)$, and $N^2 = id$, called strong negation functions, in the form $m'(x) = N_x(m(x))$ with a family $\{N_x; x \in X\}$ of such functions. This paper influenced the second, in which Enric Trillas characterized in 1979 [25] the strong negation functions by means of order-auto-morphisms φ of the unit interval as $N = \varphi^{-1} \circ (1 - id) \circ \varphi$, showing that all of them are nothing else than continuous deformations of the strong negation $1 - id$. At its turn, these two papers influenced the third, in which Sergei Ovchinnikov finally gave in 1983 [28] the general form of any (strong) negation in $[0, 1]^X$ by means of a family $\{N^X; x \in X\}$ of strong negation functions and an involution ($s^2 = id$), $s : X \to X$, provided $N_x = N_{S(X)}$ for all x in X, but without imposing the condition of keeping the complement of crisp sets.

Unfortunately, in the praxis of fuzzy sets only the negation $N = 1 - id$ is almost always used, and almost no practitioner (although not the theoreticians) never truly placed his/her attention in those three papers. It seems as if for these people there just were a unique form of negation; something that can be a bad working hypothesis in the way of *Computing with Words* [33, 34], as soon as large pieces of language should be represented and in which different forms of negation can appear, and, also, in different parts of a same large piece of language. To consider the semantics of these pieces will become strictly necessary, and, for instance, strong negations depending on points can be useful in those cases in which the negation of a predicate is not constant, but depends on the object to which the predicate is applied, and that for different objects, different forms of 'not' should be used; for instance, when the form of the negation is different in several parts of the universe of discourse, when negation is local. These forms of obtaining a fuzzy set representing 'not *P*' given that of '*P*', can be also useful for the practitioners of fuzzy logic in the task of designing a system whose linguistic description involves imprecise terms.

4.2

What could not be taken into account in those old papers since it was unknown at these times, was that although $a(P)$ and $n(P)$ are measurable if P is measurable, it is not always known which is the relation $\leq_{P'}$ that, nevertheless, is not empty by containing $\leq_P^{-1}$. That if P' is always meaningful, it is not known in general if it is effectively measurable. Without actually knowing $\leq_{P'}$ it is neither possible (strictly speaking) to specify a measure $m_{P'}$, nor a suitable negation N, either strict or strong. The problem of the effective measurability of the 'not', still remains open and for whose solution, if it exists, some linguistic examples with a known relation $\leq_{P'}$ should be submitted to a careful analysis. Indeed, the full study of the several varieties of the linguistic 'not' are yet open to study, and fuzzy practitioners remain enchained, when designing fuzzy systems, to just use those negation functions that are in the mathematical armamentarium of fuzzy logic, for instance, some strict negations, some strong negations, and N_{min}.

4.3

Concerning the complement's kernel **CP**, in the case with $X = [0, 1]$, and $P =$ big with $m_{big}(x) = x$, in which the symmetries s_{big} can be identified with the strong negations $N\varphi$, it should be noticed that m_{CP} is the inaccessible supremum of all of them that is just below the discontinuous negation N_{min}. Hence, for the predicate 'big', the complement's kernel **Cbig** can be identified with the set with only the point 0 in $[0, 1]$, that is **Cbig** $= \{0\}$, a set only allowing to assert '0 is not at all big'. At least in this case, the kernel **CP** seems to be scarcely interesting. Anyway, and since the predicate 'big' is a very simple example, more examples should be considered and, specially, with predicates whose membership functions are non-monotonic, and with symmetries not identifiable with the strong negations as it is, for instance, the predicate 'around 4' in the interval $[0, 10]$.

4.4

There is again a way that can be explored to have a crisp 'core' of negation, and one of antonymy, for a predicate P [35].

Since $m_P(x) \leq m_{P'}(x) = \varphi^{-1}(1 - \varphi(m_P(x))$, provided the negation is N_φ, this expression is equivalent with $m_P(x) \leq \varphi^{-1}(1/2) = n$, the fix point of N_φ. Hence, the set $\mathbf{N}_P = \left\{x \in X; m_P(x) \leq n\right\}$ contains those x in X showing P less than not P, and that if it is not empty and X is an interval $[a, b]$ of the real line, no doubt it exists its supremum $\alpha_P = Sup\mathbf{N}_P$, that can be viewed as similar, but not identical, to the 'separation point' advocated by Max Black; the interval $[a, \alpha_P]$ can be seen as the 'core of negation' consisting of those elements in X showing more not P than P. For instance, in the interval $[0, 10]$, with the predicate $P =$ big with membership function $m_{big}(x) = x/10$, and the negation $N(x) = 1 - x/1 + x$, for which $n = \sqrt{2} - 1$, it is $\mathbf{N}_{big} = \left\{x \in [0, 10]; x/10 \leq \sqrt{2} - 1\right\} = \left\{x \in [0, 10]; x \leq 10(\sqrt{2} - 1)\right\}$, with which it follows $\alpha_{big} = 10(\sqrt{2} - 1) \sim 4.1421$, and the 'core of negation' is the subinterval $\mathbf{N}_{big} = [0, 10(\sqrt{2} - 1)]$.

Similarly, a crisp 'core' of antonymy can be reached once an antonym aP of P is known, through the set $\mathbf{A}_P = \{x \in X; m_P(x) \le m_{aP}(x)\}$ that, since it is always $m_{aP} \le m_{P'}$, is contained in the set $\mathbf{N}_P$, provided the negation and the symmetry are coherent. Hence, provided X is an interval $[a, b]$ of the real line, it will exist a supremum β_P of this set $\mathbf{A}_P$, such that $\beta_P \le \alpha_P$, and then the core of antonymy will be included in the core of negation, $[a, \beta_P] \subseteq [a, \alpha_P]$. For instance, with 'big' in $[0, 10]$, $m_{big}(x) = x/10$, and the negation function $N(x) = 1 - x$, it is $x/10 \le 1 - x/10 \Leftrightarrow x \le 5$, and $_{big} = 5$ gives the core of negation $\mathbf{N}_P = [0, 5]$. With the symmetry $s_1(x) = 10 - x$, it is obtained the same core of antonymy, since $x/10 < (10 - x)/10 \Leftrightarrow x < 5$. Nevertheless, with the symmetry $s_2(x) = 10(1 - 10x/1 + 10x)$, from $x/10 < 10(1 - 10x/1 + 10x)/10) = 1 - x/10/1 + x/10 \Leftrightarrow x^2 + 20x - 100 < 0$, follows that the supremum $_{big}$ is $10\sqrt{2} - 10/4.1421$, and the smaller core $[0, 10\sqrt{2} - 10]$ is obtained.

It should be noticed that the core of negation of a precise predicate P in X, always coincides with its crisp complement since, in this case, and being $m_P(x) \in \{0, 1\}$, $m_{P'}(x) = 1 - m_P(x)$, is $n = 0.5$, and $m_P(x) \le 0.5$, that implies $m_P(x) = 0$. Hence, $\mathbf{N}_P = \mathbf{P}^c$. For what concerns the core of antonymy, it obviously depends on the existence of an antonym different from the negation.

Of course, if the membership function m_P is not monotonic, as it is that of 'big', the problem can be more difficult to solve. In any case, both cores depend, respectively, on the chosen strong negation function and symmetry.

4.5

It does not seem absolutely clear the existence of some reasons for which a precise predicate P should always verify the irregular coincidence $P = P'$. Let us consider a very simple example.

Take $X = [0, 10]$, and $P =$ smaller than five (shortened by **5**). It is,

$$m_5(x) = 1 \Leftrightarrow 0 \le x < 5, m_5(x) = 0 \text{ otherwise, that is, } \mathbf{P} = [0, 5),$$

and

$$m_{5'}(x) = 1 - m_5(x) = 1 \Leftrightarrow 5 \le x \le 10, m_{5'}(x) = 0 \text{ otherwise, or } \mathbf{P}^c = [5, 10],$$

corresponding to $P' =$ bigger than five. The possible antonyms $a\mathbf{5}$, will be given by membership functions

$$m_{a5}(x) = m_5(s(x)), \text{ for symmetries } s : [0, 10] \to [0, 10],$$

verifying the coherence condition $m_{a5} \le m_{5'}$. Thus, since

$$m_{a5}(x) = 1 \Leftrightarrow 0 \le s(x) < 5 \Leftrightarrow s(5) < x \le s(0), m_{a5}(x) = 0 \text{ otherwise},$$

with only that s verifies $s(5) \ne 5$ (it should be always $s(0) = 10$), $a\mathbf{5}$ will be not coincidental with **5'**. There is coincidence, for instance, with the symmetry $s_1(x) =$

$10 - x$, but it is not with the symmetry $s_2(x) = 10(10 - x/10 + x)$, with which it is $s_2(5) = 5/3$. The antonym obtained with s_2 is given by

$$m_{a5}(x) = 1 \Leftrightarrow 0 \leq 10(10 - x/10 + x) < 5 \Leftrightarrow 10/3 < x \leq 10, \text{ and } m_{a5}(x) = 0 \text{ otherwise;}$$

that is the subinterval $(10/3, 10]$ corresponding to $aP =$ bigger than $10/3$, that is different from $P' =$ bigger than five. Since $10/3 < 5$, it is 'If x is a**5**, then x is **5'**', but not reciprocally.

What seems reasonable is that, if existing in language, the antonym of a rigid predicate is, at its turn, a rigid one, but not that it should be always coincidental with the negation. Fuzzy methodology can open a door to newly looking at opposites.

4.6

To conclude, it seems that the concept of what can be seen as 'not being *P*', or 'being opposite to *P*', either from a fuzzy or a crisp point of view, still deserves more study; the subject is neither closed by what is currently known on the negation functions, either strong or just strict, nor with the use of symmetries for antonyms. There is particularly a lack of knowledge in the case *P*' is not known to be effectively measurable. When, advancing towards Zadeh's 'Computing with Words', the time of considering large linguistic phrases will arrive, and with it the necessity of a deep knowledge on the kind of linguistic separation between what 'is *P*', what 'is not *P*', and 'what is opposite to *P*', the linguistic *complements*, will appear.

Very few things are truly new in this paper, but it is perhaps in its newly structured presentation that some previously hidden hints could lie and can be found by those who, in reading it, become interested by the truly intriguing subject 'complement'.

Acknowledgments This paper is partially funded by the Foundation for the Advancement of Soft Computing (Asturias, Spain), and by the Spanish Government project MICIIN/TIN 2011-29827-C02-01.

References

1. Black, M.: Vagueness. An exercise in logical analysis. Philos. Sci. **4**, 427–455 (1937)
2. Frege, G.: Über Sinn und Bedeutung, Zeitschrift fr Philosophie und philosophische Kritik, NF 100, S. 25–50, 1892. English translation (by Black, M.): Frege, G.: On Sense and Reference, The Philosophical Review, vol. 57/3, pp. 209–230 (1948)
3. Frege, G.: Nachgelassene Schriften. Meiner, Hamburg (1969)
4. Frege, G.: Grundgesetze der Arithmetik. In: Ebert, P.A., Rossberg, M. (eds.) Jena1893–1903, Bd. II. English translation: Basic Laws of Arithmetic, Translated and Edited with an Introduction, Oxford University Press, Oxford (2013)
5. Russell, B.: Vagueness. Australas. J. Psychol. Philos. **1**, 84–92 (1923)
6. Wittgenstein, L.: Tractatus Logico-Philosophicus. Routledge & Keagan, London (1922)
7. Morris, C.W.: Foundations of the Theory of Signs, vol. 1. International Encyclopedia of Science, Chicago (1938)
8. Hempel, C.G.: Vagueness and Logic. Philos. Sci. **6**(2), 163–180 (1939)
9. Wittgenstein, L.: Philos. Investig. Blackwell Publishing, Oxford (1953)

10. Zadeh, L.A.: Quantitative fuzzy semantics. Inf. Sci. **3**, 159–176 (1971)
11. Zadeh, L.A.: Fuzzy languages and their relation to human and machine intelligence. Man and Computer. In: Proceedings of International Conference, pp. 13–165. Bordeaux, Karger, Basel (1970)
12. E. Trillas: On a Model for the Meaning of Predicates, In: Seising, R. (ed.): Views on Fuzzy Sets and Systems from Different Perspectives, Studies in Fuzziness and Soft Computing, vol. 243, pp. 175–205. Springer (2009)
13. Trillas, E., Termini, S., Moraga, C.: A naïve way of looking at fuzzy sets, forthcoming. Fuzzy Sets Syst. (2015)
14. Zadeh, L.A.: Fuzzy sets. Inf. Control **8**, 338–353 (1965)
15. Trillas, E.: Some uncertain reflections on uncertainty. Arch. Philos. Hist. Soft Comput. **I**, 1–16 (2013)
16. Kappos, D.A.: Probability Algebras and Stochastic Spaces. Academic Press, New York (1969)
17. Bodiou, G.: Théorie Dialectique des Probabilités. Gauthier-Villars, Paris (1964)
18. Trillas, E., Seising, R.: On meaning and measuring, a philosophical and historical view, forthcoming. Agora (2015)
19. Frege, G.: Begriffsschrift: eine der arithmetischen nachgebildete formelsprache des reinen Denkens. Halle a. S.: Louis Nebert, 1879. Translated as Concept Script, a formal language of pure thought modelled upon that of arithmetic, by S. Bauer-Mengelberg. In: van Heijenoort, J. (ed.) From Frege to Gödel, A Source Book in Mathematical Logic, pp. 1879–1931. Harvard University Press, Cambridge (1967)
20. Hammer, E.: Peirce's Logic, Stanford Encyclopedia of Philosophy (Fall 2010 Edition), Zalta, E.N. (ed.). http://plato.stanford.edu/entries/peirce-logic/
21. McArthur, T.: Antonym, the Oxford Companion to the English Language. Oxford University Press, Oxford (1992)
22. Trillas, E., Moraga, C., Guadarrama, S., Cubillo, S., Castiñeira, E.: Computing with Antonyms. In: Nikravesh, M., Kacprzyk, J., Zadeh, L.A. (eds.) Forging New Frontiers: Fuzzy Pioneers, Studies in Fuzziness and Soft Computing vol. I, 217, pp 133-153. Springer, Berlin (2007)
23. Trillas, E.: A model for crisp reasoning with fuzzy sets. Int. J. Intell. Syst. **27**, 859–872 (2012)
24. Halmos, P.R.: Naïve Set Theory. Springer, New York (1974)
25. Trillas, E.: Sobre funciones de negación en la teoría de los subconjuntos difusos, Stochastica, vol. III-1, pp. 47-59, 1979, (In Spanish). English version: on negation functions in fuzzy set theory. In: Barro, S., Bugarín, A., Sobrino, A. (eds.) Advances in Fuzzy Logic. Selected papers (with comments) of some Spanish Authors, Universidade de Santiago de Compostela, collecció Avances en Nr. vol. 5, pp. 31–42 (1998)
26. Bedrega, B.C.: On fuzzy negations and automorphisms. Anais do CNMAC/Brazil **2**, 1125–1131 (2009)
27. Trillas, E., Pradera, A.: Non-functional Fuzzy Connectives: The case of Negations. In: Proceedings ESTYLF (Len), pp. 527–532 (2002)
28. Ovchinnikov, S.V.: General negations in fuzzy set theory. J. Math. Anal. Appl. **92**(1), 234–239 (1983)
29. Trillas, E.: Non contradiction, excluded middle, and fuzzy sets. In: Gesu, V.D., Pal, S.K., Petrosino, A. (eds.) Proceedings 8th International Workshop, WILF 2009 Palermo, Fuzzy Logic and Applications, vol. 5571, pp 1–11. Springer, Berlin (2009)
30. Háyek, P.: Metamathematics of Fuzzy Logic. Springer, New York (1998)
31. Mas, M., Montserrat, M., Torrens, J., Trillas, E.: A survey on fuzzy implication functions. IEEE Trans. Fuzzy Syst. **15**(6), 1107–1121 (2007)
32. Lowen, R.: On fuzzy complements. Inf. Sci. **14**, 107–113 (1978)
33. Mendel, J., Zadeh, L.A., Trillas, E., Lawry, J., Hagras, H., Guadarrama, S.: What computing with words means to me? Comput. Intell. Mag. **5**(1), 20–26 (2010)
34. Zadeh, L.A.: Computing with Words. Principal Concepts and Ideas, Studies in Fuzziness and Soft Computing, vol. 277. Springer, Berlin (2012)

35. Trillas, E., García-Honrado, I.: A Layperson Reflection on Sorites. In: Seising, R., Tabacchi, M.E. (eds.) Fuzziness and Medicine: Philosophical Reflections and Application Systems in Health Care. A Companion Volume to Sadegh-Zadeh's Handbook on Analytical Philosophy of Medicine, Studies in Fuzziness and Soft Computing, vol. 302, pp 217–231. Springer, Berlin (2013)

A Note on Fuzzy-Valued Inference

Jorma K. Mattila

Abstract Fuzzy-valued inference is discussed. For that purpose, a theory of fuzzy-valued associative Kleene algebra is introduced. As an example, it is shown that fuzzy-valued Kleene algebras give a mathematical model for some fuzzy screening systems.

Keywords Fuzzy-Valued Inference · Fuzzy-Valued Kleene-Algebra · Screening Systems

1 Introduction

In fuzzy decision making, there are a lot of cases where scales of linguistic scores are used. For example, in fuzzy control linguistic expressions are generally used as control values in control processes. In fuzzy logic truth values are usually linguistic terms, like, for example, 'true', 'almost true', 'not true and not false', 'almost false', 'false'. In fuzzy screening systems linguistic values are used, too, like for example 'outstanding', 'very high', 'high', 'medium', 'low', 'very low', 'none'.

To be a scale, the scale values must have a reasonable order. For example, the truth values mentioned above are already listed in the reasonable order. Based on intuition, we ordered them using the relationship between a truth value and how near to the truth it is. The biggest difference is between truth and falsity. Similarly, the above mentioned scores for a screening system are ordered from the highest value to the lowest one. Hence, we may say that these kinds of scales are totally ordered, because the scales are finite and the ordering in any scale can be defined to be unique. However, these kinds of orderings are actually not mathematical, because they are based on intuition. Also, the use of these linguistic scores in calculations is based on intuition, even though the calculation rules can be given based on the order of the linguistic scores.

J.K. Mattila (✉)
Department of Mathematics and Physics, Lappennranta University of Technology, Skinnarilankatu 34, FI-53850 Lappeenranta, Finland
e-mail: jorma.mattila@lut.fi

M. Collan et al. (eds.), *Fuzzy Technology*, Studies in Fuzziness and Soft Computing 335, DOI 10.1007/978-3-319-26986-3_2

If we want to have a formal theory for a system of this kind, we need a mathematical counterpart to the system. Very often in many-valued logics, the truth values are given as numbers. In fuzzy systems the scores can be given as fuzzy sets. These kind of things serve the link between intuitive and formal systems. There are a lot of research about algebras for numerical truth values of many-valued logics. This research creates mathematical models for many-valued logics involved in fuzzy systems. Also, we may investigate some suitable sets of *fuzzy numbers* or fuzzy intervals in order to find some algebraic models for systems with fuzzy scores. So, we have the following basic questions:

> Does there exist some mathematical models for inference systems using scales with linguistic scores?
>
> What are the logical and mathematical bases of this kind of systems?

We give an answer to these questions in the following sections.

2 A Mathematical Background

In this presentation we introduce an algebraic approach to cases where score values are fuzzy numbers or fuzzy intervals.

We recall some earlier results for manipulating fuzzy numbers. The main things are ways of representing, ordering, and defining meets and joins of a given set of fuzzy numbers.

The representation theorem for considering fuzzy sets by means of α-cuts has been given, for example, by V. Novák [9], p. 44. A. Kaufmann and M.M. Gupta [2] (cf. pp. 19–35) consider interval arithmetics applied to triangular and trapezoidal fuzzy numbers (or fuzzy intervals) presented by means of α-cuts. They also introduced some criteria for ordering of fuzzy numbers.

Besides Kaufmaan and Gupta, also R. Fullér [1] (cf. pp. 35–36) has considered ordering of fuzzy numbers by defining *fuzzy max* and *fuzzy min* operations by means of α-cuts, and V. Novák [9] (cf. pp. 98–100) by defining *join* '$\sqcup$' and *meet* '$\sqcap$' by means of Zadeh's extension principle, as follows.

Let $\mathcal{A}$, $\mathcal{B}$ be fuzzy numbers and $x, y \in \mathbb{R}$. *Join* $\mathcal{A} \sqcup \mathcal{B}$ is the fuzzy number

$$(\mathcal{A} \sqcup \mathcal{B})(z) = \bigvee_{z=x \vee y} (\mathcal{A}(x) \wedge \mathcal{B}(y)). \tag{1}$$

Meet $\mathcal{A} \sqcap \mathcal{B}$ is the fuzzy number

$$(\mathcal{A} \sqcap \mathcal{B})(z) = \bigvee_{z=x \wedge y} (\mathcal{A}(x) \wedge \mathcal{B}(y)). \tag{2}$$

These operations appear to be the same as Fullér's *fuzzy max* and *fuzzy min*.

Novák also present the following theorem.

Theorem 1 (Novák) *The fuzzy numbers form a distributive lattice with respect to the operations '$\sqcap$' and '$\sqcup$'. It means that*

$$\begin{aligned} \mathcal{A} \sqcup \mathcal{A} &= \mathcal{A} & \mathcal{A} \sqcap \mathcal{A} &= \mathcal{A} \\ \mathcal{A} \sqcup \mathcal{B} &= \mathcal{B} \sqcup \mathcal{A} & \mathcal{A} \sqcap \mathcal{B} &= \mathcal{B} \sqcap \mathcal{A} \\ (\mathcal{A} \sqcup \mathcal{B}) \sqcup \mathcal{C} &= \mathcal{A} \sqcup (\mathcal{B} \sqcup \mathcal{C}) & (\mathcal{A} \sqcap \mathcal{B}) \sqcap \mathcal{C} &= \mathcal{A} \sqcap (\mathcal{B} \sqcap \mathcal{C}) \\ \mathcal{A} \sqcup (\mathcal{B} \sqcap \mathcal{C}) &= (\mathcal{A} \sqcup \mathcal{B}) \sqcap (\mathcal{A} \sqcup \mathcal{C}) & \mathcal{A} \sqcap (\mathcal{B} \sqcup \mathcal{C}) &= (\mathcal{A} \sqcap \mathcal{B}) \sqcup (\mathcal{A} \sqcap \mathcal{C}) \end{aligned}$$

Then he defines the ordering relation $\sqsubseteq$ for fuzzy numbers $\mathcal{A}$, $\mathcal{B}$ in the familiar way:

$$\mathcal{A} \sqsubseteq \mathcal{B} \quad \text{iff} \quad \mathcal{A} \sqcap \mathcal{B} = \mathcal{A} \quad (\mathcal{A} \sqcup \mathcal{B} = \mathcal{B} \quad \text{respectively}). \tag{3}$$

In general, fuzzy numbers do not form a linearly ordered set, except in some special cases. We will exploit some of these special cases in the following considerations.

Consider a finite set of fuzzy sets

$$T_n = \{\mathcal{A}_1, \mathcal{A}_2, \ldots, \mathcal{A}_n\} \tag{4}$$

of the interval $[0, p]$, $p \in \mathbb{R}$, $p > 0$. The fuzzy sets $\mathcal{A}_i$ $(i = 1, \ldots, n)$ of T_n satisfy the following properties:

(1°) $\mathcal{A}_i$ is either a fuzzy number or a fuzzy interval for all $1 \leq i \leq n$;
(2°) T_n is ordered, such that $\mathcal{A}_i \sqsubseteq \mathcal{A}_j$ for all $1 \leq i \leq j \leq n$;
(3°) for every $\mathcal{A}_i \in T_n$ there exists a unique fuzzy set $\neg\mathcal{A}_i \in T_n$, such that the following condition holds:

$$\neg\mathcal{A}_i = \mathcal{A}_{n-i+1}, \text{ if } 1 \leq i \leq n. \tag{5}$$

where for all $i = 1, \ldots, n$

$$\mathcal{A}_i(x) = \neg\mathcal{A}_i(p - x), \text{ if } x \in [0, p] \tag{6}$$

The operation symbol '$\neg$' is a *complementarity operation*, and we use the name *negation* for it.

The set T_n is linearly ordered, by the property (2°).

The definition of negation, i.e., the formulas (5) and (6) gives some presuppositions for the fuzzy sets in T_n. From the Eq. (6), it follows that the supports of $\mathcal{A}_i$ and $\neg\mathcal{A}_i$ satisfy the equivalency

$$\text{supp}\mathcal{A}_i = [a, b] \quad \Longleftrightarrow \quad \text{supp}\neg\mathcal{A}_i = [p - b, p - a] \tag{7}$$

There exist two special cases, namely $\alpha = 0$ and $\alpha = 1$. These conditions give the closures of the support and core of a given fuzzy set $\mathcal{A}$. Now, because

$[\mathcal{A}_i]^0 = \mathrm{cl}(\mathrm{supp}\,\mathcal{A}_i)$ and $[\mathcal{A}_i]^1 = \mathrm{cl}(\mathrm{core}\,\mathcal{A}_i)$ then the closures of the supports and the cores of $\mathcal{A}_i$ and $\neg\mathcal{A}_i$ satisfy the equivalencies

$$\mathrm{cl}(\mathrm{supp}\,\mathcal{A}_i) = [a_0, b_0] \iff \mathrm{cl}(\mathrm{supp}\,\neg\mathcal{A}_i) = [p - b_0, p - a_0] \tag{8}$$

$$\mathrm{cl}(\mathrm{core}\,\mathcal{A}_i) = [a_1, b_1] \iff \mathrm{cl}(\mathrm{core}\,\neg\mathcal{A}_i) = [p - b_1, p - a_1] \tag{9}$$

The lengths of the intervals $[\mathcal{A}_i]^\alpha$ and $[\neg\mathcal{A}_i]^\alpha$ are the same for any α, because if $[\mathcal{A}_i]^\alpha = [a_\alpha, b_\alpha]$ then its length is $b_\alpha - a_\alpha$, and the length of $[\neg\mathcal{A}_i]^\alpha$ is $p - a_\alpha - (p - b_\alpha) = b_\alpha - a_\alpha$, too.

By means of these considerations above, it is easy to see that the fuzzy sets $\mathcal{A}_i$ and $\neg\mathcal{A}_i$ are symmetric with respect to the vertical line $x = \frac{p}{2}$. The value $\frac{p}{2}$ is the centre of the interval $[0, p]$.

Our next purpose is to show that the set T_n (see (4)) forms a *Kleene algebra* of fuzzy numbers belonging to the set T_n. First, we have to show that the set T_n forms a DeMorgan algebra. About DeMorgan algebras, see Rasiowa [10]. (Rasiowa uses the name *quasi-Boolean algebra* for DeMorgan algebra.) To do this, we prove the following Lemmas.

Lemma 1 *The system $\mathcal{T}_n = \langle T_n, \sqcup, \sqcap\rangle$ is a distributive and complete lattice.*

Proof $\mathcal{T}_n$ is a distributive lattice by means of Theorem 1. It is also complete because for any two elements $\mathcal{A}_i, \mathcal{A}_j \in T_n$ $(1 \le i, j \le n)$ the expressions $\mathcal{A}_i \sqcup \mathcal{A}_j$ and $\mathcal{A}_i \sqcap \mathcal{A}_j$ are defined and T_n is closed under the operations $\sqcup$ and $\sqcap$, i.e., $\mathcal{A}_i \sqcup \mathcal{A}_j$ equals to either $\mathcal{A}_i$ or $\mathcal{A}_j$, and $\mathcal{A}_i \sqcap \mathcal{A}_j$ equals to either $\mathcal{A}_j$ or $\mathcal{A}_i$ respectively. □

Lemma 2 *The law of double negation*

$$\neg\neg\mathcal{A}_i = \mathcal{A}_i \tag{10}$$

for any $\mathcal{A}_i \in T_n$ holds in the lattice $\mathcal{T}_n$.

Proof The result follows from the formula (5) by an easy calculation. □

Lemma 3 *De Morgan Laws hold on T_n.*

Proof Let $\mathcal{A}_i, \mathcal{A}_j \in T_n$ be any two elements, such that $\mathcal{A}_i \sqsubseteq \mathcal{A}_j$. Hence, $i \le j$, by the property (2°). Further, $\neg\mathcal{A}_i = \mathcal{A}_{n-i+1}$ and $\neg\mathcal{A}_j = \mathcal{A}_{n-j+1}$, by (5). Comparing the subindices $n - i + 1$ and $n - j + 1$ we see that $n - j + 1 \le n - i + 1$ because $i \le j$, by assumption. Hence, $\mathcal{A}_{n-j+1} \sqsubseteq \mathcal{A}_{n-i+1}$, i.e., $\neg\mathcal{A}_j \sqsubseteq \neg\mathcal{A}_i$. So, the implication $\mathcal{A}_i \sqsubseteq \mathcal{A}_j \Longrightarrow \neg\mathcal{A}_j \sqsubseteq \neg\mathcal{A}_i$ holds. It is easy to see that this implication holds to the other direction, too. Hence, the equivalency

$$\mathcal{A}_i \sqsubseteq \mathcal{A}_j \iff \neg\mathcal{A}_j \sqsubseteq \neg\mathcal{A}_i \tag{11}$$

holds for any $\mathcal{A}_i, \mathcal{A}_j \in T_n$. Further, $\neg\mathcal{A}_j \sqcup \neg\mathcal{A}_i = \neg\mathcal{A}_i$, by (3), and hence,

$$\neg(\neg\mathcal{A}_j \sqcup \neg\mathcal{A}_i) = \neg\neg\mathcal{A}_i = \mathcal{A}_i = \mathcal{A}_i \sqcap \mathcal{A}_j$$

by assumption and by (10). Hence, one of De Morgan Laws,

$$\mathcal{A}_i \sqcap \mathcal{A}_j = \neg(\neg\mathcal{A}_j \sqcup \neg\mathcal{A}_i) \tag{12}$$

holds. The other De Morgan law, $\mathcal{A}_i \sqcup \mathcal{A}_j = \neg(\neg\mathcal{A}_j \sqcap \neg\mathcal{A}_i)$, follows from (12) by replacing $\mathcal{A}_i$ and $\mathcal{A}_j$ with $\neg\mathcal{A}_i$ and $\neg\mathcal{A}_j$, respectively, and applying the law of double negation. □

From the Lemmas 1, 2 and 3 it follows that the system $\mathcal{T}_n = \langle T_n, \sqcup, \sqcap, \neg, \mathcal{A}_n \rangle$ is *De Morgan algebra*, because T_n is a non-empty set, $\mathcal{T}_n = \langle T_n, \sqcup, \sqcap \rangle$ is a distributive lattice with top element $\mathcal{A}_n$, $\neg$ is a unary operation on T_n, and $\mathcal{T}_n$ satisfies the law of double negation and De Morgan laws.

The top and bottom elements exist in $\mathcal{T}_n$ because $\mathcal{T}_n$ is finite totally ordered set. Now we also know that the complementarity $\neg$ is *quasi-complementation*. (See closer considerations, for example, in Rasiowa [10], p. 44–45.) The top element $\mathcal{A}_n$ is the neutral element of the operation $\sqcap$. In $\mathcal{T}_n$, there exists a bottom element, too, namely $\mathcal{A}_1$, which is the neutral element of the operation $\sqcup$. Especially, by the definition of negation, the conditions $\neg\mathcal{A}_1 = \mathcal{A}_n$ and $\neg\mathcal{A}_n = \mathcal{A}_1$ hold in $\mathcal{T}_n$. It is a general case that any De Morgan algebra has top element and bottom element being the negations of each other.

If a De Morgan algebra satisfies so-called *Kleene condition*, it is a *Kleene algebra*. So, the last thing before getting a fuzzy-valued Kleene algebra is to check whether the Kleene condition

$$\mathcal{A}_i \sqcap \neg\mathcal{A}_i \sqsubseteq \mathcal{A}_j \sqcup \neg\mathcal{A}_j\,, \text{ if } 1 \leq i, j \leq n \tag{K}$$

holds in our De Morgan algebra $\mathcal{T}_n = \langle T_n, \sqcup, \sqcap, \neg, \mathcal{A}_n \rangle$. Here the condition (K) is constructed for lattices where the elements are fuzzy numbers or fuzzy intervals.

Theorem 2 *The algebra* $\mathcal{T}_n = \langle T_n, \sqcup, \sqcap, \neg, \mathcal{A}_n \rangle$ *is a Kleene algebra.*

Proof The algebra $\mathcal{T}_n$ is De Morgan algebra, as we have noticed above. So, we have to show that the algebra $\mathcal{T}_n$ satisfies the Kleene condition (K).

Let $\mathcal{A}_i, \mathcal{A}_j \in T_n$ be arbitrarily chosen, hence $\neg\mathcal{A}_i, \neg\mathcal{A}_j \in T_n$, too, because T_n is closed under negation. Suppose $\mathcal{A}_i \sqsubseteq \mathcal{A}_j$ whenever $i \leq j$, for all $\mathcal{A}_i, \mathcal{A}_j \in T_n$.

If the number of fuzzy sets in T_n is $n = 2k + 1$ (i.e., n is odd) then the middle element of T_n is $\mathcal{A}_{k+1}$, and hence $\neg\mathcal{A}_{k+1} = \mathcal{A}_{k+1}$, by the definition of negation.

If $n = 2k$ (i.e., n is even) then $\neg\mathcal{A}_k = \mathcal{A}_{k+1}$ and $\neg\mathcal{A}_{k+1} = \mathcal{A}_k$, by the definition of negation.

We denote

$$\mathcal{A}_{[\frac{n}{2}]} = \begin{cases} \mathcal{A}_{k+1} & \text{if} \quad n = 2k+1 \\ \mathcal{A}_k & \text{if} \quad n = 2k \end{cases}$$

Hence, for any $\mathcal{A}_i$, if $\mathcal{A}_i \sqsubseteq \mathcal{A}_{[\frac{n}{2}]}$ then $\mathcal{A}_{[\frac{n}{2}]} \sqsubseteq \neg\mathcal{A}_i$, and vice versa. Hence, for any i, $\mathcal{A}_i \sqcap \neg\mathcal{A}_i \sqsubseteq \mathcal{A}_{[\frac{n}{2}]}$ and for any j, $\mathcal{A}_{[\frac{n}{2}]} \sqsubseteq \mathcal{A}_j \sqcap \neg\mathcal{A}_j$. This completes the proof. □

Especially, $\mathcal{T}_n = \langle T_k, \sqcup, \sqcap, \neg, \mathcal{A}_n \rangle$ is an associative Kleene algebra, by Theorem 1.

3 Construction of Applicable Kleene Algebras

Examples about easily manipulable fuzzy sets in applications based on Kleene algebras of fuzzy sets are triangular fuzzy numbers, Gaussian fuzzy numbers, other bell-shaped fuzzy numbers and fuzzy intervals.

As an example, consider a trapezoidal fuzzy interval

$$\mathcal{A}(x) = \begin{cases} 0 & \text{if } x < a_1 \\ \frac{x-a_1}{a_2-a_1} & \text{if } a_1 \leq x < a_2 \\ 1 & \text{if } a_2 \leq x < a_3 \\ \frac{a_4-x}{a_4-a_3} & \text{if } a_3 \leq x \leq a_4 \\ 0 & \text{if } x > a_4 \end{cases} \tag{13}$$

on a closed real interval $[0, p]$ $(0 < p,\ p \in \mathbb{R})$ and $x, a_1, a_2, a_3, a_4 \in [0, p]$ and $0 \leq a_1 < a_2 < a_3 < a_4$. If $a_2 = a_3$ then $\mathcal{A}$ is a triangular fuzzy number. The support and the core of $\mathcal{A}(x)$ are the intervals $[a_1, a_4]$ and $[a_2, a_3]$, respectively. The increasing part on the left and the decreasing part on the right side of $\mathcal{A}$ have the membership functions

$$\mathcal{A}_L(x) = \begin{cases} 0 & \text{if } x < a_1, a_2 < x \\ \frac{x-a_1}{a_2-a_1} & \text{if } a_1 \leq x < a_2 \end{cases}, \quad \mathcal{A}_R(x) = \begin{cases} \frac{a_4-x}{a_4-a_3} & \text{if } a_3 \leq x \leq a_4 \\ 0 & \text{if } x < a_3, a_4 < x \end{cases} \tag{14}$$

If the supports $\text{supp}(\mathcal{A}_L) = [a_1, a_2]$ and $\text{supp}(\mathcal{A}_R) = [a_3, a_4]$ have the same size then the membership function of the fuzzy interval $\mathcal{A}$ is symmetric with respect to the vertical line $x = \frac{a_3+a_2}{2}$.

Applying, for example, some considerations, due to Novák [9] and Kaufmann and Gupta [2], $\mathcal{A}$ can be considered as an ordered 4-tuple $\mathcal{A} = (a_1, a_2, a_3, a_4)$ where $\mathcal{A}(x)$ is increasing if $x \in [a_1, a_2]$, $\mathcal{A}(x) = 1$ if $x \in [a_2, a_3]$, and $\mathcal{A}(x)$ is decreasing if $x \in [a_3, a_4]$. Further, the negation for $\mathcal{A}$ on the interval $[0, p]$ can be given as the ordered 4-tuple

$$\neg\mathcal{A} = (p - a_4, p - a_3, p - a_2, p - a_1) \tag{15}$$

which is a fuzzy interval on $[0, p]$ of x-axis. This fuzzy interval can be given in the form, similar to $\mathcal{A}$ in (13), as

$$\neg\mathcal{A}(x) = \begin{cases} 0 & \text{if } x < p - a_1 \\ \frac{x-p+a_3}{a_3-a_2} & \text{if } p - a_3 \leq x < p - a_2 \\ 1 & \text{if } p - a_3 \leq x < p - a_2 \\ \frac{p-a_1-x}{a_2-a_1} & \text{if } p - a_2 \leq x \leq p - a_1 \\ 0 & \text{if } x > p - a_1 \end{cases} \tag{16}$$

Consider a set of fuzzy trapezoidal intervals $T_n = \{\mathcal{A}_1, \mathcal{A}_2, \ldots, \mathcal{A}_n\}$ of the interval $[0, p]$ where the *divisional points* of $[0, p]$ on x-axis, determined by the fuzzy intervals $\mathcal{A}_i$ $(i = 0, \ldots, n)$ are $a_0, a_1, \ldots, a_k$, where $k = 2n - 1$. The fuzzy intervals

$$\mathcal{A}_1(x) = \begin{cases} 1 & \text{if } 0 \leq x < a_1 \\ \frac{a_2-x}{a_2-a_1} & \text{if } a_1 \leq x \leq a_2 \\ 0 & \text{if } x > a_2 \end{cases}, \quad \mathcal{A}_n(x) = \begin{cases} 0 & \text{if } x < a_{k-2} \\ \frac{x-a_{k-2}}{a_{k-1}-a_{k-2}} & \text{if } a_{k-2} \leq x < a_{k-1} \\ 1 & \text{if } a_{k-1} \leq x < a_k \end{cases} \tag{17}$$

are the first and the last fuzzy interval, respectively, which can be given by means of the 4-tuples

$$\mathcal{A}_1 = (0, 0, a_1, a_2) \quad \text{and} \quad \mathcal{A}_n = (a_{k-2}, a_{k-1}, a_k, a_k).$$

Here we agree that $\mathcal{A}_1(x) = 1$ if $x \in [0, a_1]$ and $\mathcal{A}_n(x) = 1$ if $x \in [a_{k-1}, a_k]$. The other fuzzy intervals between $\mathcal{A}_1$ and $\mathcal{A}_k$ can be given in the form

$$\mathcal{A}_2 = (a_1, a_2, a_3, a_4),\ \mathcal{A}_3 = (a_3, a_4, a_5, a_6),\ \ldots,\ \mathcal{A}_{n-1} = (a_{k-4}, a_{k-3}, a_{k-2}, a_{k-1}).$$

Note that the first divisional point on the interval $[0, p]$ is origo, and the last one is $a_k = p$. Hence, $\mathcal{A}_1(0) = 1$ and $\mathcal{A}_n(a_k) = \mathcal{A}_n(p) = 1$.

Especially, on x-axis, the divisional points for the intervals being parts of the supports of the fuzzy sets in T_n are as follows. The number of fuzzy sets in T_n is n and the number of the divisional points is $2n$. So, we have the divisional points $a_0, a_1, a_2, \ldots, a_k$, where $k = 2n - 1$. And the first divisional point is $a_0 = 0$ and the last one $a_k = a_{2n-1} = p$.

The supports of all the fuzzy numbers form a cover to the interval $[0, p]$, such that the union of the covers of $\mathcal{A}_1$, $\mathcal{A}_2$ …, $\mathcal{A}_n$ is exactly the same as the interval itself, i.e.,

$$[0, p] = \bigcup_{i=1}^{n} \operatorname{supp} \mathcal{A}_i. \tag{18}$$

Example 1 Consider a collection of fuzzy intervals on $[0, p]$ where the fuzzy intervals are of the form (16) and (17) where $n = 7$. Hence, the set T_7 consists of the fuzzy intervals

$$\mathcal{A}_1 = (0, 0, a_1, a_2),\ \mathcal{A}_2 = (a_1, a_2, a_3, a_4),\ \mathcal{A}_3 = (a_3, a_4, a_5, a_6),$$
$$\mathcal{A}_4 = (a_5, a_6, a_7, a_8),\ \mathcal{A}_5 = (a_7, a_8, a_9, a_{10}),\ \mathcal{A}_6 = (a_9, a_{10}, a_{11}, a_{12}),$$
$$\text{and } \mathcal{A}_7 = (a_{11}, a_{12}, a_{13}, a_{13})$$

i.e.,

$$T_7 = \{\mathcal{A}_1, \mathcal{A}_2, \mathcal{A}_3, \mathcal{A}_4, \mathcal{A}_5, \mathcal{A}_6, \mathcal{A}_7\}$$

The divisional points are

$$a_0 = 0, a_1, a_2, a_3, a_4, a_5, a_6, a_7, a_8, a_9, a_{10}, a_{11}, a_{12}, a_{13} = p.$$

The algebra $\mathcal{T}_7 = \langle T_7, \sqcup, \sqcap, \neg, \mathcal{A}_7 \rangle$ satisfies all the properties considered in Sect. 2.

4 An Application Example: Fuzzy Screening Systems

As a motivating example, we consider a case of a *fuzzy screening system*, a technique suggested by R. Yager (for example, see Yager [11]). These systems contain fuzzy data. The source material for this description about fuzzy screening systems in this section is taken from Robert Fullér's book [1] *Introduction to Neuro-Fuzzy Systems*.

A fuzzy screening system is a two stage process as follows:

- In the first stage, experts are asked to provide an evaluation of the alternatives. This evaluation consists of a rating for each alternative on each of the criteria.
- In the second stage, the methodology is used to aggregate the individual experts evaluations to obtain an overall linguistic value for each object.

The problem consists of three components.

(1) The first component is a collection

$$X = \{X_1, \ldots, X_p\}$$

of *alternative solutions* from amongst which we desire to select some subset to be investigated further.

(2) The second component is a group

$$A = \{A_1, \ldots, A_r\}$$

of *experts* whose opinion solicited in screening the alternatives.

Table 1 Scale of scores

Score name	Label	Score symbol
Outstanding	OU	S_7
Very high	VH	S_6
High	H	S_5
Medium	M	S_4
Low	L	S_3
Very low	VL	S_2
None	N	S_1

(3) The third component is a collection

$$C = \{C_1, \dots, C_m\}$$

of *criteria* which are considered relevant in the choice of the objects to be further considered.

For each alternative, each expert is required to provide his/her opinion. In particular, for each alternative an expert is asked to evaluate how well that alternative satisfies each of the criteria in the set C. These evaluations of alternative satisfaction to criteria will be given in terms of elements from the scale S in Table 1.

Based on intuition, the use of such a scale provides a natural ordering, $S_i > S_j$, if $i > j$, and the maximum and minimum of any two scores be defined by

$$\max\{S_i, S_j\} = S_i\,, \quad \text{if } S_i \geq S_j \tag{19}$$

$$\min\{S_i, S_j\} = S_j\,, \quad \text{if } S_i \leq S_j. \tag{20}$$

where max and min are fuzzy max and min defined in [1], i.e., these operations are the same as $\sqcup$ and $\sqcap$, respectively. Using our notation above, these conditions can be expressed in the form

$$S_i \sqcup S_j = S_j\,, \quad \text{if} \quad S_i \sqsubseteq S_j \tag{21}$$

$$S_i \sqcap S_j = S_i\,, \quad \text{if} \quad S_i \sqsubseteq S_j. \tag{22}$$

Thus for an alternative an expert provides a collection of n values

$$\{P_1, \dots, P_n\},$$

where P_j is the rating of the alternative on the jth criterion by the expert. Each P_j is an element in the set of allowable scores S,

$$S = \{S_7, S_6, S_5, S_4, S_3, S_2, S_1\}.$$

Assuming $n = 6$, an example of a typical scoring for an alternative from one expert would be

$$\{S_4, S_3, S_7, S_6, S_7, S_1\}$$

or, using the labels of the scores,

$$\{\mathrm{M}, \mathrm{L}, \mathrm{OU}, \mathrm{VH}, \mathrm{OU}, \mathrm{N}\}.$$

Independent of this evaluation of alternative satisfaction to criteria, each expert must assign a measure of importance to each of the criteria. An expert uses the same scale, S, to provide the importance associated with the criteria.

The next step in the process is to find the overall evaluation for an alternative by a given expert. For this we use a methodology suggested by Yager [11]. A crucial aspect of this approach is the taking of the negation of the importances as

$$\mathrm{Neg}(S_i) = S_{q-i+1} \qquad (q \text{ is the number the scores in } S).$$

For the scale S the negation operation provides the following:

$$\mathrm{Neg}(\mathrm{OU}) = \mathrm{N}, \mathrm{Neg}(\mathrm{VH}) = \mathrm{VL}, \mathrm{Neg}(\mathrm{H}) = \mathrm{L}, \mathrm{Neg}(\mathrm{M}) = \mathrm{M},$$
$$\mathrm{Neg}(\mathrm{L}) = \mathrm{H}, \mathrm{Neg}(\mathrm{VL}) = \mathrm{VH}, \mathrm{Neg}(\mathrm{N}) = \mathrm{OU}.$$

Then the unit score of each alternative by each expert, denoted by U, is calculated as follows:

$$U = \min_j \{\mathrm{Neg}(I_j) \vee P_j\}, \tag{23}$$

where I_j denotes the importance of the jth criterion.

Because in our algebra $\neg I_j \sqcup P_j$ is the same as $\mathrm{Neg}(I_j) \vee P_j$ in Yager's system then, using the notation of our algebra, the unit score formula is

$$U = (\neg I_1 \sqcup P_1) \sqcap \ldots \sqcap (\neg I_m \sqcup P_m) \tag{24}$$

If we think that the operations of the screening systems are logical connectives, we note that the unit score formula (23) essentially is an *anding* of the criteria satisfactions modified by the importance of the criteria. The formula (23) can be seen as a measure of the degree to which an alternative satisfies the following statement:

All important criteria are satisfied.

The following example is considered the case where the experts $A_1, \ldots, A_r$ evaluates his/her opinion about the importance of each criterion C_i (i = $1, \ldots, 5$) by using the scores from the scale S. Then the experts evaluate how well each alternative X_i ($i = 1 \ldots, p$) satisfies each criterion. So, for example, an alternative X_i gets the evaluation given in Table 2 from an expert A_j.

Table 2 Evaluations for one alternative by one expert

Criterion	C_1	C_2	C_3	C_4	C_5
Importance	VH	VH	M	L	VL
Satisfaction score	M	L	OU	VH	OU

Example 2 Consider an alternative X_i with the following scores on five criteria in the following Table 2. An expert A_j gives his/her scores to the importance of each criterion and his/her scores to the alternative, how well the alternative meets each criterion.

Using this evaluation, we apply the truth value evaluation based on the Kleene algebra $\mathcal{K}_7 = \langle S, \sqcap, \sqcup, \neg, S_7 \rangle$, and the unit evaluation for the alternative X_i from the expert A_j is

$$\begin{aligned} U_{ij} &= (\mathrm{VL} \sqcup \mathrm{M}) \sqcap (\mathrm{VL} \sqcup \mathrm{L}) \sqcap (\mathrm{M} \sqcup \mathrm{OU}) \sqcap (\mathrm{H} \sqcup \mathrm{VH}) \sqcap (\mathrm{VH} \sqcup \mathrm{OU}) \\ &= \mathrm{M} \sqcap \mathrm{L} \sqcap \mathrm{OU} \sqcap \mathrm{VH} \sqcap \mathrm{OU} = \mathrm{L}. \end{aligned} \tag{25}$$

We note that comparing this result with that in the original example[1] with the same data we see that the results are identical.

The essential reason for the low performance of this objects is that it performed low on the second criterion which has a very high importance. Linguistically, Eq. (23) is saying that *If a criterion is important then an alternative should score well on it.*

The satisfaction scores S_i $(i = 1, \ldots, 7)$ are interpreted by fuzzy numbers or fuzzy intervals, for example, by the same fuzzy intervals as in Example 1 in Sect. 3, such that the scale of satisfaction scores

$$S = \{S_1, S_2, S_3, S_4, S_5, S_6, S_7\} \tag{26}$$

forms an associative Kleene algebra $\mathcal{K}_7 = \langle S, \sqcup, \sqcap, \neg, S_7 \rangle$. So, the calculation tools of these inferences are based on this algebra. Hence, fuzzy screening systems serve as a practical application example about fuzzy-valued Kleene algebras.

As a result of the first stage, we have for each alternative X_i a collection of evaluations

$$\{P_{i1}, \ldots, P_{ir}\}, \quad i = 1, \ldots, p \tag{27}$$

where $P_{ik} \in S$ is the unit evaluation of the ith alternative by the kth expert.

Example 2 shows how to aggregate the individual experts evaluations in order to get an overall linguistic value for each object after the evaluations made by each expert.

---

[1] See Robert Fullér [1], Ex. 1.18.1., p. 111–112.

5 Concluding Remarks

As a conclusion, we can answer in the affirmative to the questions we stated in the end of Sect. 1.

The construction of the above presented Kleene algebra is a mathematical basis for applications where fuzzy quantities are used as fuzzy scores in fuzzy inferences, like fuzzy screening systems. Hence, we found a mathematical model for fuzzy screening systems.

It must be noted that this kind of mathematical models work only if the linguistic scores correspond to the fuzzy quantities, i.e., fuzzy numbers or fuzzy intervals, in corresponding algebras. The linguistic interpretations of the fuzzy quantities are subjective, i.e., they are based on personal opinions. Hence, we cannot totally get rid of intuition. However, this thing is a strength in fuzzy systems, if we use it carefully.

Kleene algebras have a central role in fuzzy set theory. The author has shown that Prof. Zadeh's theory of standard fuzzy sets he presented in [12] is based on Kleene algebras. First, in symposium "Fuzziness in Finland", 2004, and in the paper [4] the author showed that Zadeh's theory in [12] forms a De Morgan algebra, and later on, for example, in [7] it is shown that standard fuzzy set theory forms a Kleene algebra. The author is calling this algebra by name *Zadeh algebra*.

Kleene algebras for fuzzy quantities serve an algebraic basis for *many-fuzzy-valued logics*. Truth values of this kind of logics are fuzzy numbers. The author has some ideas and sketches to create some fuzzy-many-valued logical systems. Some preliminaries are already considered in Mattila [5, 6].

References

1. Fullér, R.: Introduction to Neuro-Fuzzy Systems. Advances in Soft Computing. Physica-Verlag, Heidelberg (2000)
2. Kaufmann, A., Gupta, M.M.: Fuzzy Mathematical Models in Engineering and Management Science. North-Holland, New York (1988)
3. Lowen, R.: Fuzzy Set Theory. Basic Concepts, Techniques and Bibliography. Kluwer Academic Publishers, Dordrecht (1996)
4. Mattila, J.K.: Zadeh algebras as a syntactical approach to fuzzy sets. In: Baets, D., Caluwe, D., Fodor, D.T., Zadrożny, K. (eds.) Current Issues in Data and Knowledge Engineering, Problemy Współczesnej Nauki Teoria I Zastosowania, Informatyka, Akademicka Oficyna Wydawnicza EXIT, Warszawa 2004, (Selected papers presented at EUROFUSE'2004, Warszawa, Poland on September 22-25, 2004), pp. 343–349 (2004)
5. Mattila, J.K.: On fuzzy-valued propositional logic. Walden, P., Fullér, R., Carlsson, J. (eds.) Expanding the Limits of the Possible, p. 33– 43. ISBN 952-12-1817-7, Åbo (2006)
6. Mattila, J.K.: Standard fuzzy sets and some many-valued logics. Dadios, E.P. (ed.) Fuzzy Logic—Algorithms, Techniques and Implementations. In:Tech. ISBN 979-953-51-0393-6, pp. 75–96 (2012)
7. Mattila, J.K.: Zadeh algebra as a basis of Łukasiewicz logics. In: Proceedings of NAFIPS 2012 Meeting, I978-1-4673-2338-3/12/31.00 (2012)
8. Negoită, C.V., Ralescu, D.A.: Applications of Fuzzy Sets to Systems Analysis. Birkhäuser (1975)

9. Novák, V.: Fuzzy Sets and Their Applications. Adam Hilger, Philadelphia (1989)
10. Rasiowa, H.: An Algebraic Approach to Non-Classical Logics. North-Holland, New York (1974)
11. Yager, R.R.: Fuzzy screening systems. In: Lowen, R., Roubens, M. (eds.) Fuzzy Logic: State of the Art, pp. 251–261. Kluwer, Dordrecht (1993)
12. Zadeh, L.A.: Fuzzy sets. Inf. Control **8** (1965)

A Concept Map Approach to Approximate Reasoning with Fuzzy Extended Logic

Vesa A. Niskanen

Abstract Concept maps and Lotfi Zadeh's fuzzy extended logic are applied to such computerized approximate reasoning models as modus ponens and modus tollens. A statistical application is also sketched. A pedagogical approach is mainly adopted, but these ideas are also applicable to the conduct of inquiry in general.

1 Introduction

Approximate reasoning with fuzzy systems has already proven to be a plausible method in various applications. In particular, Lotfi Zadeh's recent studies on establishing the principles of the extended fuzzy logic, FLe, which is a combination of "traditional" provable and "precisiated" fuzzy logic, FLp, as well as a novel meta-level "unprecisiated" fuzzy logic, Flu, opens various interesting prospects [31–38]. He states that in the FLp the objects of discourse and analysis can be imprecise, uncertain, unreliable, incomplete or partially true, whereas the results of reasoning, deduction and computation are expected to be provably valid. In the Flu, in turn, the membership functions and generalized constraints are not specified, and they are a matter of perception rather than measurement. In addition, in the FLp we use precise theorems, classical deducibility and formal logic, whereas the FLu operates with informal and approximate reasoning [36]. The FLe stems from Zadeh's previous theories on information granulation, precisiated language and computing with words, as well as on the theory of perceptions [32–35, 37, 38].

The Author has applied the ideas of the FLe to theory formation, hypothesis verification and scientific explanation, inter alia [21–23]. These examinations, in turn, are based on semantic validity of reasoning, and within the FLe this means

V.A. Niskanen (✉)
Department of Economics and Management, University of Helsinki, Helsinki, Finland
e-mail: vesa.a.niskanen@helsinki.fi

M. Collan et al. (eds.), *Fuzzy Technology*, Studies in Fuzziness and Soft Computing 335, DOI 10.1007/978-3-319-26986-3_3

that, instead of traditional p-validity, we use approximate reasoning and other approximate entities, the f-entities. These f-entities are approximate counterparts of the corresponding traditional constructions. For example, approximate theories are approximate counterparts of the traditional theories [21].

However, the FLe approach is unable to provide us the big picture on the complex phenomena of the real world in which everything usually depends upon everything else. In the scientific theories and models this means that we have to operate in such networks of variables in which various interconnections prevail between them. Robert Axelrod and Bart Kosko have provided a partial resolution within time series analysis to this problem with their ideas on the cognitive maps, and quite many fuzzy applications are already available in this area [2, 7, 8, 15, 16, 18, 27]. The well-known examples of other applications are the Bayesian networks, theory of networks, structural equation modelling, answer tree analysis, and even factor analysis [5, 39].

Below we apply the concept maps when we examine the complex phenomena, and this theory mainly stems from the ideas of Joseph Novak and certain pedagogical theories [1, 24]. These maps resemble the cognitive maps, but their application areas are more diverse. Thanks for the concept maps, we may consider and understand complex phenomena quite conveniently, and in particular the students may find this type of modelling useful in their learning processes. Today we meet various challenges in teaching because novel learning theories and tools are available. Examples of these are deep learning, e-learning and intelligent learning paths and materials.

To date the use of traditional concept maps has mainly based on manual work and a priori models. We will, in turn, consider how the concept maps may be used in a computer environment when the interrelationships between the variables or concepts are specified with the FLe and approximate reasoning. We also consider a posteriori models. In this manner we may construct and simulate complex models by mimicking the human reasoning with computers, and hence we may utilize better these maps. Our approach may enhance theory formation and model construction in the conduct of inquiry as well as provide viable tools for future learning environments for students.

2 Cognitive and Concept Maps

Cognitive and concept maps enable us to understand and examine complex phenomena of the real world. Their computerized models, in turn, open new prospects for understanding, simulating and forecasting these phenomena. Fuzzy systems are already quite widely applied to the cognitive maps, whereas the concept maps are still a quite new frontier. Below we consider first the essential features of these maps.

2.1 Cognitive Maps

The fuzzy numeric cognitive maps stem from the ideas of Robert Axelrod and Bart Kosko [2, 16], and they are used for simulating and forecasting such phenomena of the real world that consist of numerous variables and their interrelationships. They may also include feedback operations, and hence in these systems everything may depend upon everything else. We usually apply these maps to such simulations in which we aim to forecast the complex phenomena on the time axis [7, 8, 15, 18, 27]. In statistics the structural equation models are used for this purpose (e.g., Mplus™, LISREL™, AMOS™) as well as time series analysis, but the fuzzy cognitive maps are usually simpler and more robust in model construction.

The traditional Axelrod's cognitive maps base on classical bivalent or trivalent logic and mathematics and hence they can only model more or less roughly the relationships. Kosko enhanced these maps by using variable values that usually range from 0 to 1 and the corresponding values denote the degrees of activation of the variables. The degrees of relationship between the variables, in turn, may range from -1 to 1 in which case the benchmarks -1, 1 and 0 denote full negative effect, full positive effect and no effect, respectively. Due to the mathematical properties of the numeric cognitive maps, in iterations the values of its variables oscillate, are chaotic or finally become stable.

If (empiric) data in a given period of time is unavailable, we only operate with a priori maps, and thus we only use human intuition or expertise in our constructions, otherwise we can also construct a posteriori maps and then we apply such methods as statistics (e.g., regression and path analysis), neural networks or evolutionary computing. Hence, appropriate data as well as such reasoning as abduction or induction can yield usable cognitive maps in an automatic manner.

However, the numeric cognitive maps can only establish monotonic causal interrelationships between the variables, whereas fuzzy linguistic cognitive maps enable us to avoid this problem [7, 29]. The latter approach is also more user-friendly due to its linguistic nature.

In the linguistic maps we can assign fuzzy linguistic values to our variables and establish their interrelationships with linguistic expressions by using fuzzy linguistic rule bases. Hence, we may operate with both quantitative and qualitative linguistic variables, use more diverse variable values and interrelationships, and apply nonlinear modeling. On the other hand, we are usually unable to apply them if we operate with such dynamical systems that include a great number of variables.

2.2 Concept Maps

If we replace the cognitive maps with the concept maps, we may apply model construction in a more diverse manner because then the time-dependent models are not our only options. The concept maps generally enable us to consider and

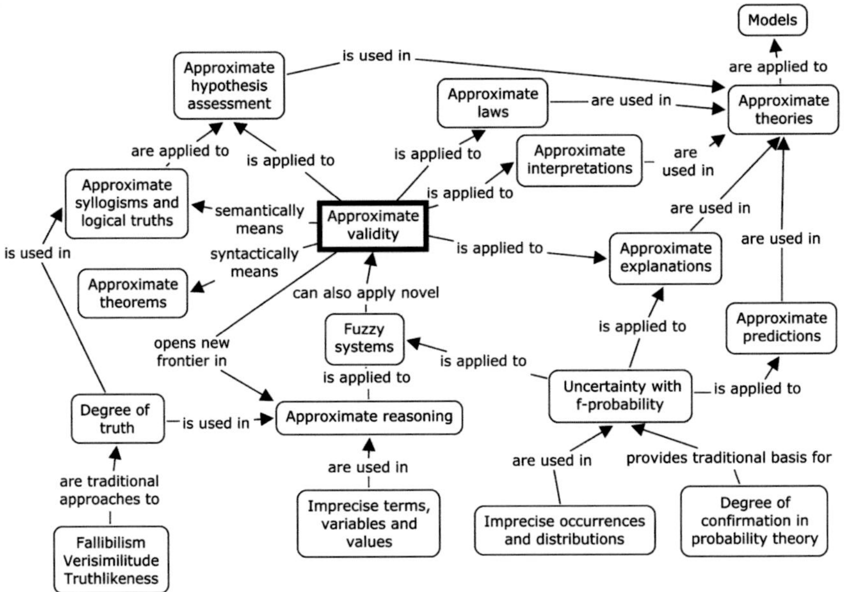

Fig. 1 A simple concept map on approximate validity within the FLe

understand in a holistic manner the complex phenomena of the real world when we specify the concepts or variables and their interrelationships in a given context or framework [24] (Fig. 1). Hence, they may be applied to such tasks as problem-setting, brainstorming, transformation of tacit knowledge, concept analysis, simulation, decision-making, learning and even for therapeutic purposes when we aim to represent our expertise, knowledge or data in a given context.

Effective and meaningful learning is one central application area, and we will focus quite much on this issue below. In brief, from the standpoint of the behavioral sciences, the human learning comprises cognitive, affective and psycho-motoric factors. Their objects are facts, emotions and attitudes, and skills, respectively. The meaningful learning attempts to give a broader perspective to the real world and, unlike only applying the traditional passive rote learning, the pupil or student is encouraged to be an active agent who is creative, seeks for new information, is able to process the acquired information and can create his/her personal conceptual constructions or frameworks. Today the cutting-edge learning methods also include the diverse use of e-learning, information and communication technology, learning games and simulations, intelligent individual learning paths and even intelligent learning materials.

The concept maps aim to enhance and provide tools for meaningful cognitive learning, and in this respect they stem from the learning theory known as constructivism as well as from Ausubel's assimilation theory [1]. However, these maps in their traditional form still lack their computerized models. Within fuzzy systems some models are already available but they base on various background theories (e.g., [3, 17, 28]). We apply the FLe because it provides us with a comprehensive theory on approximate reasoning.

Sometimes there is only a thin red line between the cognitive and concept maps, or they are even interchangeable. Below we construct concept maps by applying fuzzy systems and approximate reasoning, and then we may examine our constructions at theoretical level and even run practical simulations with these systems, if necessary. These models may thus open novel prospects for examining complex phenomena in learning and even in the conduct of inquiry.

First we consider certain central aspects of approximate reasoning by using concept maps. These maps illuminate well these tasks even for the beginners, but we may also use their frameworks in our deeper methodological studies and computer simulations.

3 Truth Valuation with Concept Maps

In approximate reasoning we should fulfill two conditions in truth valuation. First, when scientific realism is adopted, our conception on truth should correspond well with natural language, human reasoning and the real world. Second, we should formulate simple and plausible mathematical and logical operations with fuzzy systems for them. Below we consider these aspects by adopting a concept map approach with linguistic variables and relationships.

It seems that in truth valuation the fuzzy community usually applies either explicitly or implicitly the correspondence theory of truth, and Alfred Tarski suggested the well-known definition that applied this idea [9, 10],

$$\text{the expression 'x is P' is true, iff x is P (in the real world)} \quad (1)$$

Hence, this means that truth manifests the relationship between the linguistic and real world. For example, the linguistic expression 'Christer Carlsson lives in Turku' is true if and only if Christer lives in Turku (the Swedish name for Turku is Åbo, both are used in Finland). Since this definition draws a distinction between the object and meta languages, and thus avoids such classic paradoxes as the Liar, we will also apply this idea below.

Within fuzzy systems and approximate reasoning we still seem to have some bivalent commitments, and one of these problems is related to the exegesis on antonyms and negations of expressions. In approximate reasoning we may usually draw a clear distinction between these concepts, and hence, for example, such expressions as 'nontrue' and 'false' or 'nonyoung' and 'old' have distinct meanings, and we should bear this in mind when specifying our logical operations. Below we assume that 'nontrue' and 'nonfalse' mean anything but true and false, respectively.

The Author has also applied the idea used in the theory of truthlikeness, in which case we evaluate the truth vales of the basic or primitive expressions in the light of their true counterparts [20–23]. Hence, we in fact apply fuzzy similarity relation to

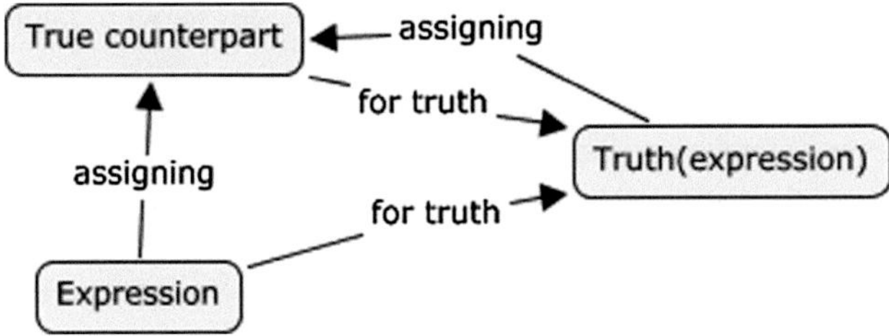

Fig. 2 A concept map on assigning values for modified expressions

these expressions, and our truth values range from maximum (true) to minimum (false) degrees of similarity. For example, the truth value of 'John's age is 25 years', provided that he is actually 22, i.e., Truth(John is 25 // John is 22), is close to true (high degree of similarity). Figure 2 depicts a simple concept map that provides a basis for our truth valuation, and it also depicts the idea on finding other values in a given context. Generally, if the values of two nodes are fixed, we may obtain the third value (see also below).

Hence, we may now establish our basic metarules for truth valuation [23]:

$$\begin{aligned}
&1.\ \text{'x is P' is true, iff x is P}\\
&2.\ \text{'x is P' is false, iff x is the antonym of P}\\
&3.\ \text{'x is P' is not true, iff x is not P}\\
&4.\ \text{'x is P' is not false, iff x is not the antonym of P}
\end{aligned} \tag{2}$$

For example,

1. 'John is young' is true, iff John is young
2. 'John is young' is false, iff John is old
3. 'John is young' is not true, iff John is not young
4. 'John is young' is not false, iff X is not old

If we are unable to assign an appropriate antonym for the expression P, we may use its negation. For the linguistic modifiers, such as 'very' and 'fairly', we usually use Osgood's or Likert's scales because they are successfully applied to the human sciences. The former scale is applied below and we use odd number of linguistic values, for example,

$$\text{P, fairly P, neither P nor Q (neutral or middle value), fairly Q, Q,}$$

in which P and Q are antonyms. This idea also concerns our truth values in which case our antonyms are 'false' and 'true'.

Table 1 presents now our meta rules for specifying the truth values for these expressions, and these may also be used as the central fuzzy rules when the

Table 1 Fuzzy meta rules for assigning truth values with basic expressions

	True counterpart					
Expression		**P**	**Fairly p**	**Neither p nor q**	**Fairly q**	**Q**
	P	True	Fairly true	Half true	Fairly false	False
	Fairly p	Fairly true	True	Fairly true	Half true	Fairly false
	Neither p nor q	Half true	Fairly true	True	Fairly true	Half true
	Fairly q	Fairly false	Half true	Fairly true	True	Fairly true
	Q	False	Fairly false	Half true	Fairly true	True

corresponding fuzzy reasoning system is constructed ('half true' means neither false nor true). For example, given the antonyms 'young' and 'old',

$$\text{Truth(John is young //John is fairly young)} = \text{fairly true.}$$

When operating with fuzzy systems in a computer environment, we have to specify the corresponding fuzzy sets, and within the truth values we use such sets as depicted in Fig. 3 that are in fact fuzzy numbers from zero (false) to unity (true) [4]. We may also apply this "horizontal" method for the other values if the range of the reference set is assigned properly.

Hence, we may construct a fuzzy reasoning system that applies such prevailing reasoning as Mamdani or Takagi-Sugeno reasoning for truth valuation, and Figs. 4 and 5 depict examples on a rule set and fitting based on this idea.

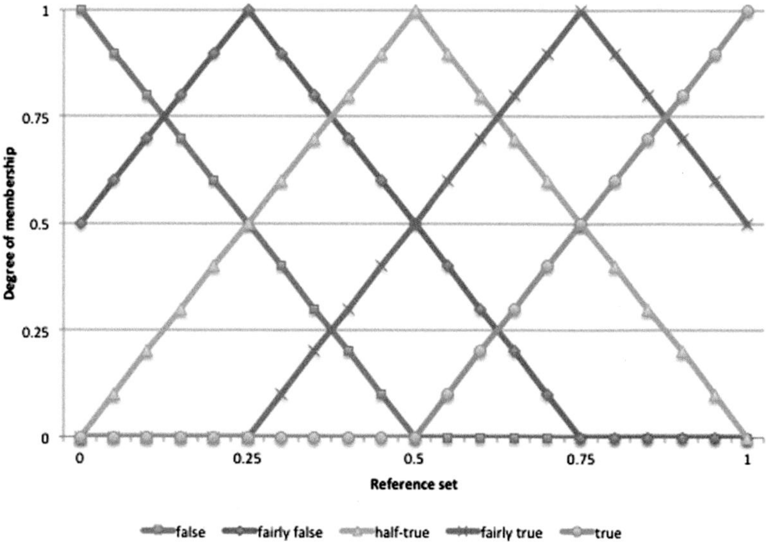

Fig. 3 Examples of fuzzy sets that represent our basic truth values

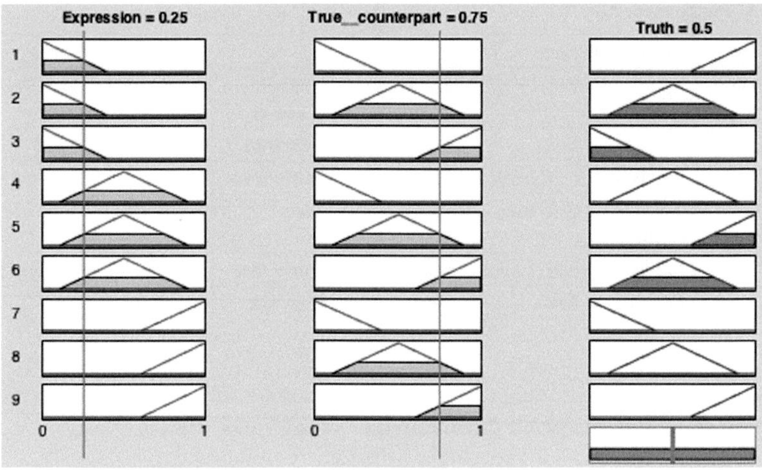

Fig. 4 An example of a fuzzy rule set for assigning the basic truth values when an expression and its true counterpart are given

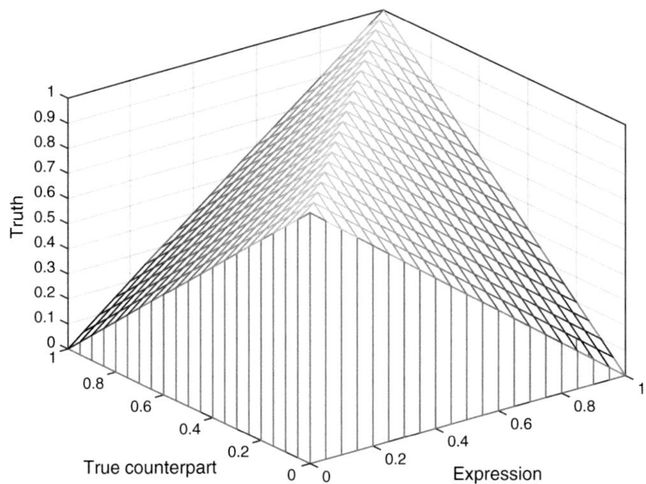

Fig. 5 An example of a fuzzy reasoning model fitting for assigning the basic truth values

Given an expression and its truth value, we may also apply indirectly the rules in Table 1 when we assign the corresponding true counterpart. For example, if Truth (John is fairly young) = fairly true, our fuzzy rules yield that John is actually either young or middle-aged. Table 2 presents these rules directly.

In the case of negation Table 1 also yields indirectly plausible truth values in the manner of if we assume that 'not P' means anything but P. For example,

Table 2 Fuzzy meta rules for assigning true counterparts with basic expressions

	Truth value					
Expression		**False**	**Fairly false**	**Half true**	**Fairly true**	**True**
	P	Q	Fairly Q	Neither P nor Q	Fairly P	P
	Fairly P		Q	Fairly Q	P or neither P nor Q	Fairly P
	Neither P nor Q			P or Q	Fairly P or fairly Q	Neither P nor Q
	Fairly Q		P	Fairly P	Q or neither P nor Q	Fairly Q
	Q	P	Fairly P	Neither P nor Q	Fairly Q	Q

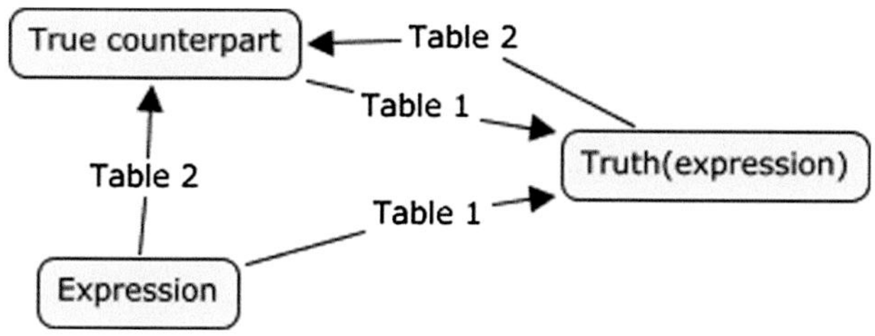

Fig. 6 A concept map with fuzzy rules on assigning values for modified expressions

$$\text{Truth(not P //P)} = \text{ from fairly true to false} = \text{not true} \tag{3}$$

Hence, Fig. 6 depicts a modified version of our original concept map that applies our considerations.

We may now apply the concept map in Fig. 6 as flowchart to simple computer simulations within truth evaluations. These simulations are illuminating especially for pedagogical purposes.

We may also construct concept maps for the compound expressions and quantifiers with the foregoing methods but these considerations are precluded. Fuzzy meta rules for these relationships are suggested in [23].

4 Approximate Syllogisms with Concept Maps

The modus ponens and modus tollens syllogisms are widely used in reasoning and the conduct of inquiry. The approximate version of the former also plays an essential role in fuzzy reasoning, whereas the latter is less studied even though it opens interesting prospects for approximate hypothesis verification. The

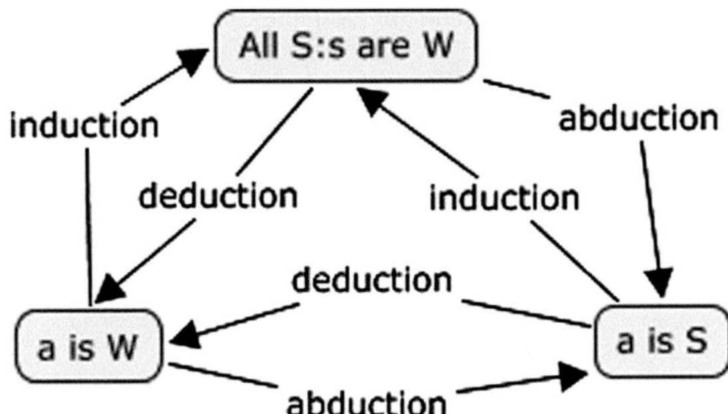

Fig. 7 Concept map on the three basic types of relationships between the premises and the conclusion

approximate versions of these syllogisms also gain various advantages over their traditional counterparts, and these aspects are considered below.

At a general level in reasoning, we may thus start from the framework depicted in Fig. 7, if we focus on the relationship between the premises and the conclusion. [10, 19, 25, 26]. Hence, according to Charles Peirce, we may perform deduction, induction or even abduction [19].

For example, given the classic example by Aristotle [19],

$$\begin{aligned}&1.\,\text{All swans are white, (All S are W)},\\&2.\,\text{a is a swan, (a is S)},\\&3.\,\text{a is white, (a is W)},\end{aligned} \tag{4}$$

in our framework the reasoning from the items 1 and 2 to 3 apply deduction, from 2 and 3 to 1 induction, and from 1 and 3 to 2 abduction. Our syllogisms enable us to consider all these approaches.

4.1 Modus Ponens

With an approximate modus ponens, we may apply the framework depicted in Fig. 8, when the truth values of implication are established. From the logical standpoint, we usually presuppose that the implication is true whenever the truth value of its antecedent is less than or equal with its consequent. Below we also adopt this approach, and in fact, we apply the linguistic version of the well-known Lukasiewicz's implication,

$$\text{Truth}(\text{if A, then B}) = \min(\text{true}, \text{true} - \text{Truth}(\text{A}) + \text{Truth}(\text{B})), \tag{5}$$

when our truth values denote the above-mentioned fuzzy numbers. It also holds that the greater the positive difference between the truth values of the consequent and antecedent, the lower is the truth value of implication (Table 3). We may also construct the corresponding fuzzy rule-based system (Figs. 9 and 10).

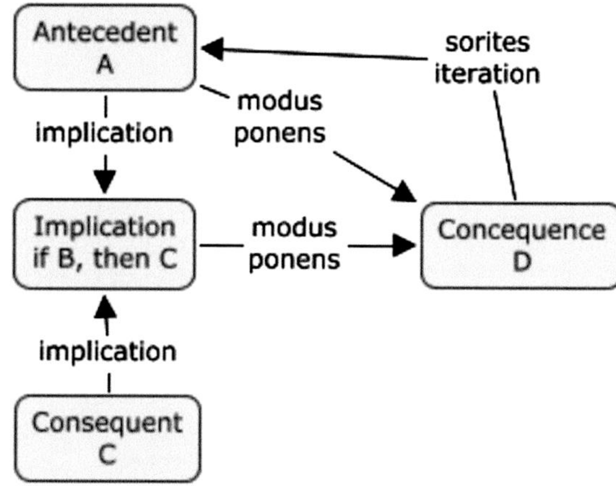

Fig. 8 A concept map on the general framework for an approximate modus ponens

Table 3 Fuzzy meta rules for assigning the truth values for implication

	Consequent					
Antecedent		**False**	**Fairly false**	**Half true**	**Hairly true**	**True**
	False	True	True	True	True	True
	Fairly false	Fairly true	True	True	True	True
	Half true	Half true	Fairly true	True	True	True
	Fairly true	Fairly false	Half true	Fairly true	True	True
	True	False	Fairly false	Half true	Fairly true	True

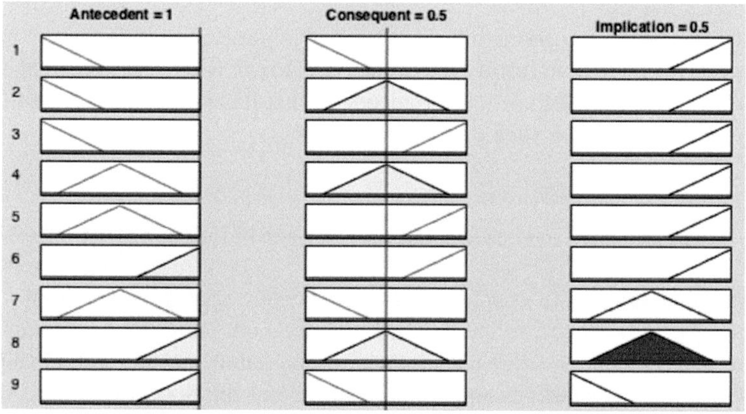

Fig. 9 An example of a fuzzy rule set for assigning the truth values for implication

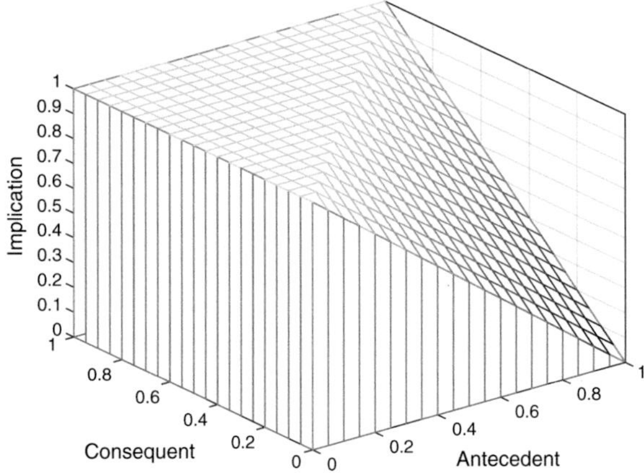

Fig. 10 An example of a fuzzy-reasoning model fitting for assigning the truth values of implication

Our concept map also depicts the approximate modus ponens reasoning,

$$\begin{aligned}&\text{A,}\\&\text{If B, then C,}\\&\text{Hence, D}\end{aligned} \tag{6}$$

in which A and C are more or less similar to B and D, respectively, and in the bivalent case A = B and C = D. For example, given that Truth(A) = fairly true and the implication is true, our meta rules in Table 3 yield Truth(D) = from fairly true to true.

Since we have also taken into account the nontrue implications, our modus ponens may be applied to resolve the Sorites paradox in which case the implication is nontrue, but it is nevertheless recommendable that it is more or less close to true. As we know, given now such premises as

$$\begin{aligned}&\text{A person aged 20 is young (true),}\\&\text{If a person aged 20 is young, then a person aged 21 is young (} < \text{true),}\end{aligned}$$

our reasoning draws the conclusion Truth(a person aged 21 is young) < true (Table 4, Fig. 11). Hence, in the next iteration the both premises are nontrue, and the conclusion will be less true than in the previous round, and so forth. Finally, we only obtain false conclusions and we may thus avoid the Sorites paradox.

Once again, the relationships in our concept map may base on our fuzzy rules in Table 4, and then we may perform simulations.

Table 4 Fuzzy meta rules for assigning the truth values for modus ponens reasoning

	Implication					
Antecedent		**False**	**Fairly false**	**Half true**	**Fairly true**	**True**
	False					False to true
	Fairly false				False	Fairly false to true
	Half true			False	Fairly false	Half true to true
	Fairly true		False	Fairly false	Half true	Fairly true to true
	True	False	Fairly false	Half true	Fairly true	True

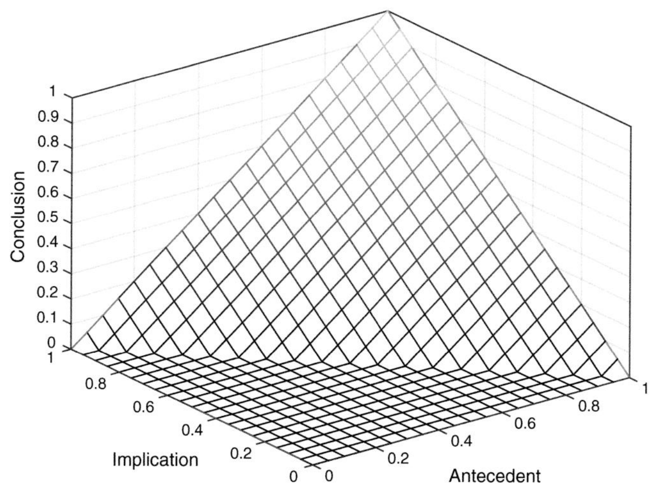

Fig. 11 An example of a fuzzy-reasoning model fitting for assigning the truth values for the conclusions in modus ponens

Figure 12 depicts one simulation for our Sorites reasoning. In this context a fuzzy cognitive map may also be used.

Here again our original concept map may be modified when the fuzzy relationships are included in it (Fig. 13).

Modus ponens is already widely used within approximate reasoning, whereas our next syllogism, the modus tollens, still expects further examination. Below we consider its role in hypothesis verification by also taking into account the pedagogical aspects.

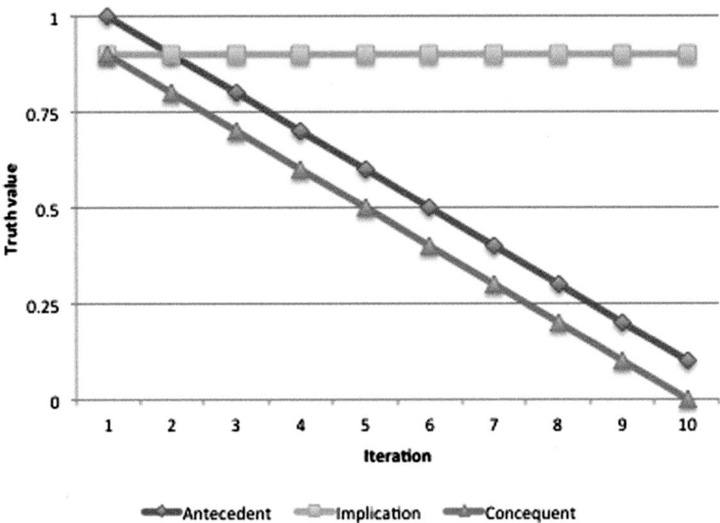

Fig. 12 An example of approximate modus ponens reasoning when implication is nontrue

Fig. 13 A concept map with fuzzy rules on the general framework for an approximate modus ponens

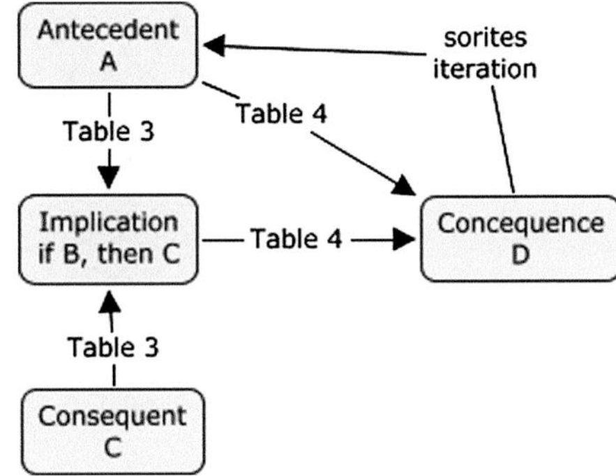

4.2 Modus Tollens

The bivalent version of the modus tollens is in the form,

$$\begin{aligned}&\text{If A, then B,}\\&\text{not B,}\\&\text{Hence, not A.}\end{aligned} \tag{7}$$

In modus ponens the antecedent of implication usually determines the conclusion, whereas now the consequent plays an essential role. However, since in the bivalent reasoning the negation of an expression and its antonym are usually identical concepts, we should modify this syllogism for our purposes. Hence, its approximate form might be [22],

$$\begin{aligned}&\text{If A, then B,}\\&\text{C,}\\&\text{Hence, D,}\end{aligned}\tag{8}$$

in which C and D are more or less similar to B and A, respectively (Fig. 14). If C is now the antonym of B, we conclude that D is the antonym of A, and this reasoning is analogous to the classic bivalent case (Table 5, Fig. 15).

As in the case of the approximate modus ponens, we may also operate here with nontrue implications in which case we have more dispersion or fuzziness in our conclusions. Table 5 also provides guidelines for this type of reasoning, but in practice the truth values close to true only seem plausible for implication. Figure 16, in turn, depicts our concept map with fuzzy relationships.

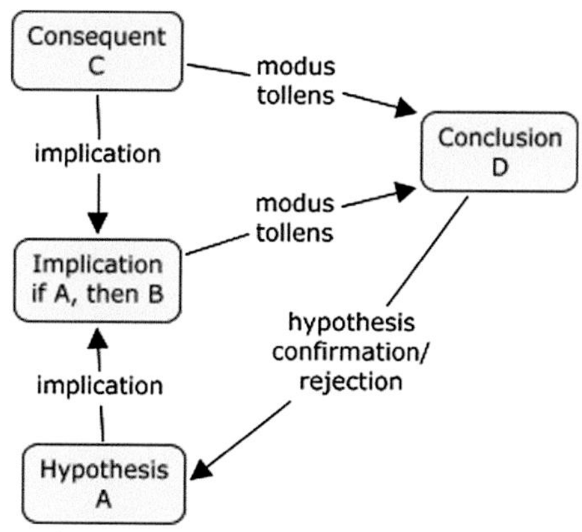

Fig. 14 A concept map on the general framework for an approximate modus tollens

Table 5 Fuzzy meta rules for assigning the truth values for modus tollens reasoning

	Implication					
Consequent		**False**	**Fairly false**	**Half true**	**Fairly true**	**True**
	False	True	Fairly true	Half true	Fairly false	False
	Fairly false		True	Fairly true	Half true	False to fairly false
	Half true			True	Fairly true	False to half true
	Fairly true				True	False to fairly true
	True					False to true

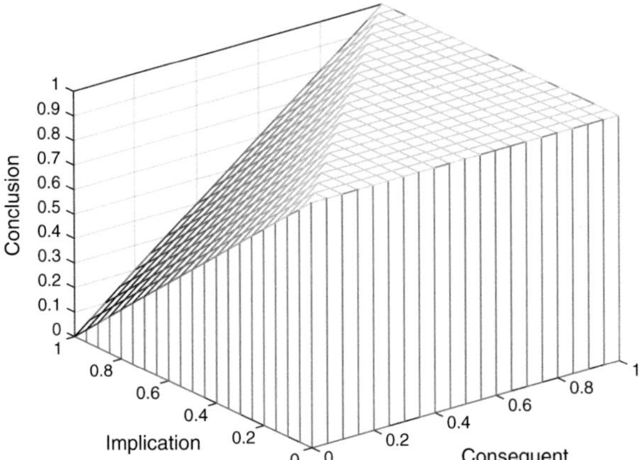

Fig. 15 An example of a fuzzy reasoning truth-value fitting for modus tollens inference

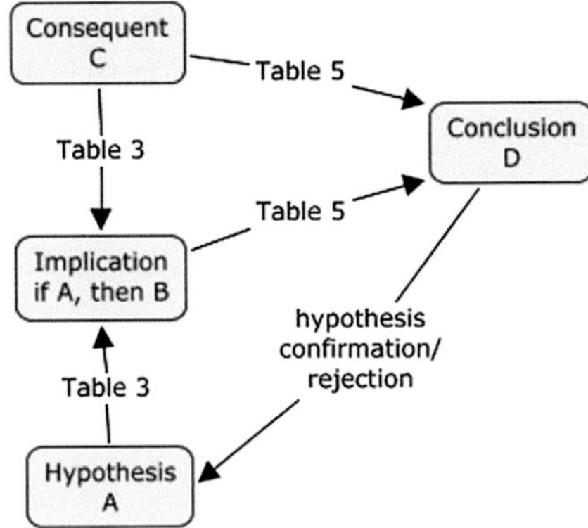

Fig. 16 A concept map with fuzzy rules on the general framework for an approximate modus tollens

The modus tollens, in turn, plays an essential role in the hypothesis verification, in particular within the widely-used hypothetico-deductive method [11, 12, 19, 22, 25, 26] (Fig. 17). From the logical standpoint, the hypotheses may be singular, general or probabilistic statements, i.e., they can refer to singular objects or occurrences, regularities or more or less uncertain events. Hence, they may deal with observational statements but also with the phenomena at a general level or even include purely theoretical concepts.

To date, the hypothesis verification has mainly based on bivalent reasoning, but we can also apply the FLe and our approximate reasoning in this context. In fact, the conventional hypothesis verification in the quantitative research usually comprises three reasoning methods, John Stuart Mill's method of difference, the disjunctive syllogism and the modus tollens syllogism [19].

The hypothetico-deductive method with the traditional modus tollens assumes that we first establish our hypothesis, H, in a given test or experiment, and then we deduce the effects or consequences, C, according to H. If C does not correspond to

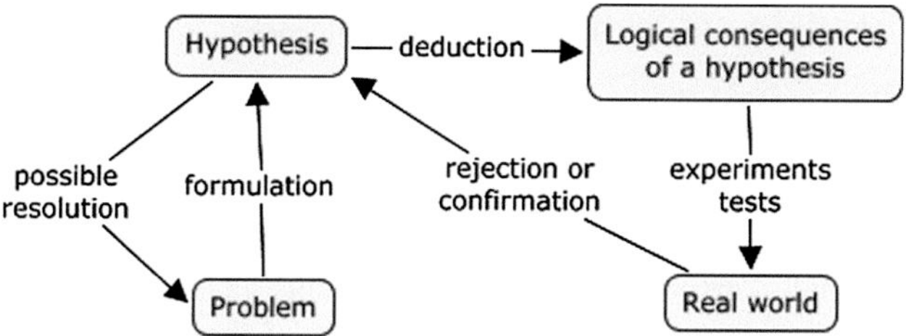

Fig. 17 A concept map on the general framework for hypothesis verification with hypothetico-deductive reasoning

our evidence, such as theories or observations, we will reject our hypothesis. If, in turn, C is derived by using induction or probabilistic methods, this approach is known as hypothetico-inductive method.

For example, when deduction is performed, given the hypothesis that Christer is 10 years old,

$$\begin{array}{l}\text{If Christer is 10, then he is still a bachelor. (deduction)}\\ \text{Christer is married. (evidence)}\\ \text{Hence, our hypothesis is false and we reject it.}\end{array} \quad (9)$$

If, on the other hand, C corresponds to our evidence, we are unable to apply (6), and then one resolution is only to consider the degree of confirmation of H, i.e., in this context we actually apply induction [19, 25, 26]:

$$\begin{array}{l}\text{If Christer is 10, then he is still at school.}\\ \text{Christer is a schoolboy.}\\ \text{Hence, our hypothesis is more or less confirmed.}\end{array} \quad (10)$$

This idea, in turn, is closely related to such approach to epistemic probability in which we assess the degree of confirmation according to the relationship between the hypothesis and its evidence [6, 13, 14, 30]. By applying the FLe, we may adopt this approach in our approximate reasoning. Hence, given the implication, if H, then C, and the truth of C, we obtain the truth of H (Table 5). If C is false, H is also false, and this is analogous to the bivalent case. On the other hand. the closer C is to truth, the closer H is also to truth, in other words, the higher is the degree of confirmation of our hypothesis. It follows that,

$$\begin{array}{l}\text{If Christer is 10, then he is still a bachelor. (true implication)}\\ \text{Christer is married. (Truth(Christer is a bachelor) = false according to the evidence)}\\ \text{Hence, Truth(John is 10) = false, and we reject this hypothesis.}\end{array} \quad (11)$$

If Christer is 10, then he is still at school. (true implication)
Christer is a schoolboy. (Truth(Christer is a schoolboy) = true according to the evidence)
Hence, Truth(Christer is 10) $\geq$ false, and our hypothesis is more or less confirmed.
(12)

Since we usually carry out several experiments or tests for accepting our hypotheses, our final resolutions are aggregations of these procedures. The two widely-used presuppositions are that, first, in our experiments the obtained degrees of confirmation should be sufficiently high, and, second, our experiments should increase the previous probabilities or degrees of confirmation of our hypotheses [13, 14, 19]. The former criterion expresses the idea on high degree of confirmation, and the latter is known as positive relevance.

In our approach the former criterion might be that our hypothesis should be at least half true in each experiment. Hence, those experiments that yield low truth values to hypotheses will either be ignored or they will decrease our final degree of confirmation. The latter option is more recommendable in the conduct of inquiry.

The positive relevance may also be specified when an appropriate aggregation operation for the degrees of confirmation, DC, are used. For example, given the DC according to the hypothesis H and evidence E, $DC(H, E)_{now}$, in our present experiment and $DC(H)_{total}$ based on the aggregation of our previous experiments, the traditional methods basically assess the difference $DC(H, E)_{now} - DC(H)_{total}$ that should be positive [13, 14, 19]. We may also apply this idea by specifying the new aggregation value, $DC(H)_{newtotal}$,

$$DC(H)_{newtotal} = \text{Aggregation}(DC(H)_{total}, DC(H, E)_{now}). \tag{13}$$

The acceptance of our hypothesis then depends upon the final value of DC $(H)_{total}$, and this procedure is analogous to the statistical tests when the levels of significance are calculated.

Figure 18 depicts an example on 20 more or less successful experiments by applying iteration to the concept map in Fig. 16. In this case the cumulative mean was used for aggregation because it is sufficiently simple for demonstrating our idea,

$$DC(H)_{newtotal} = \text{Mean}(DC(H)_i, DC(H, E)_{now}),\ i = 1, 2, \ldots, n, \tag{14}$$

in which $DC(H)_1, \ldots, DC(H)_n$ refer to the aggregations of the previous experiments. If the values of $DC(H)_i$ are monotonically increasing, we maintain the idea on the positive relevance. A fuzzy reasoning system may also replace (14). In Fig. 18 the positive relevance was not fulfilled.

In practice we also aim at maximally informative hypotheses even though the higher the truth value or probability of a hypothesis, the lower its semantic information content, and vice versa [13]. For example, given the hypotheses,

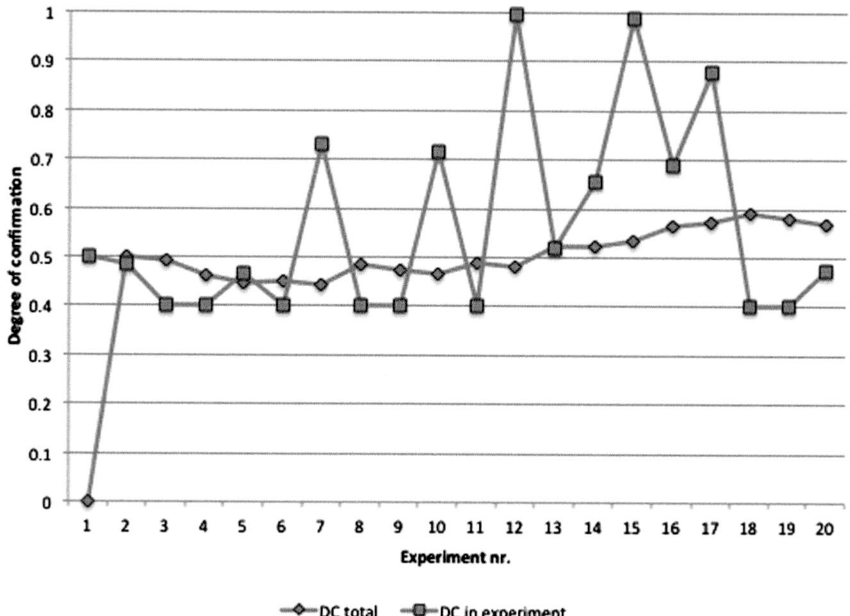

Fig. 18 An example on simulation when the final degree of confirmation of a hypothesis is assessed according to 20 experiments

Christer Carlsson lives in Turku.

Christer Carlsson lives in Europe.

The former has higher information content because it excludes more alternatives than the latter, whereas the latter is more likely true. Hence, in this sense the scientists are "gambling with truth" in hypothesis verification. In our framework this means that the positively relevant experiments also increase the information content.

The foregoing considerations have more or less orientated to pedagogical aspects, but below we adopt a more scientific approach when we apply the concept maps and approximate reasoning to statistical analysis.

5 Concept Maps in Statistical Modeling

We may also apply concept maps to statistical analysis when we study the interrelationships between the variables. Hence, we may, to some extent at least, replace such conventional methods as regression analysis, canonical correlation analysis, structural equation modeling and answer tree analysis with our approximate reasoning models in a given concept map. In time series analysis we may also use the cognitive maps.

As a simple example, we consider briefly below the Voter data that is included in the SPSS™ statistical software. This data contains 1847 observations on the Presidential election in the USA in 1992. We use four node variables, viz.,

1. Voting: respondent's vote for Presidential election (Bush, Perot or Clinton).
2. Age: respondent's age in years.
3. Education (degree): respondent's highest degree (0 = less than high school, 1 = high school, 2 = junior college, 3 = Bachelor or 4 = graduate degree).
4. Gender.

The concept map in Fig. 19 depicts some conventional analyses that may be carried out in this context [39]. If approximate reasoning is applied instead, we may specify these interrelationships with fuzzy rule sets and the corresponding fuzzy reasoning models. We adopt both of these approaches below.

First, the contingency table tests (e.g., Fisher's test, Goodman's and Kruskal's tau, Cramer's V, Bonferroni tests) showed that there was no relationship (connection) between gender and education.

Second, the contingency table tests with gender and voting seem predict that the males principally favored Bush or Perot and females supported Clinton (Fig. 20).

In a fuzzy reasoning model for this relationship we may thus use such rules as

1. If gender was male, then the vote fairly likely went for Bush or Perot.
2. If gender was female, then the vote fairly likely went for Clinton.

Third, the respondent's education also seemed to affect on his/her voting behavior to some extent when the contingency table was analyzed (Fig. 21).

With the fuzzy rules this meant that

1. If the highest degree was from high school to Bachelor's degree, the vote fairly likely went for Bush or Perot.
2. If the highest degree was less than high school or graduate degree, the vote fairly likely went for Clinton.

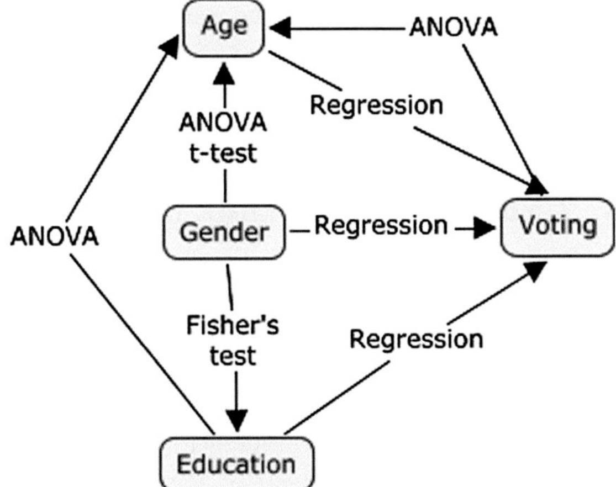

Fig. 19 Some interrelationships according to the Voter data

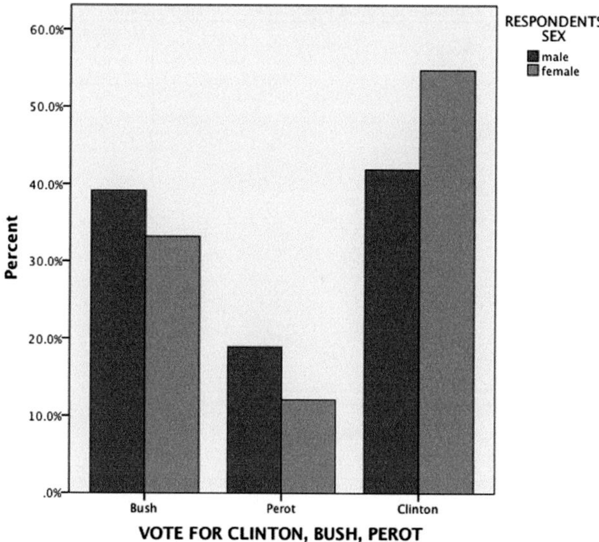

Fig. 20 The voting behavior according to gender in the Voter data

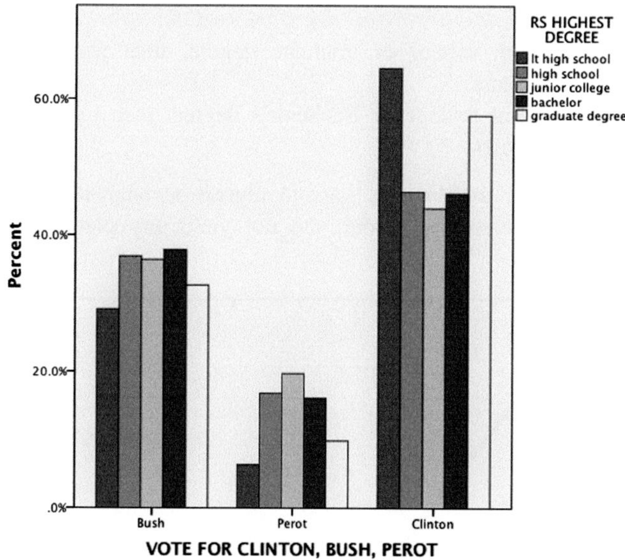

Fig. 21 The voting behavior according to education in the Voter data

Fourth, according to the tests of the two-way ANOVA with age as the test variable and education and gender as the factors, we may conclude that sex has no effect and the interaction term a slight effect at most, whereas between the degrees

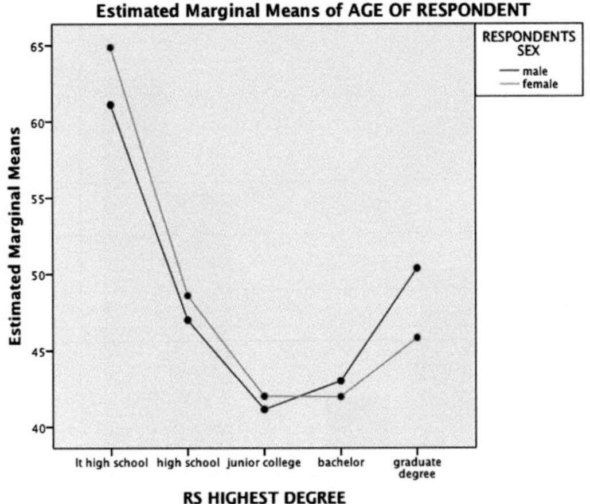

Fig. 22 Group means of age among gender and the degrees of education with two-way ANOVA

on education some differences in ages are found (Figs. 22 and 23). The corresponding fuzzy rules for this relationship would be,

1. If education is less than high school, then respondent's age is approximately 63 years.
2. If education is high school or graduate degree, then respondent's age is approximately 48 years.
3. If education is junior college or Bachelor's degree, then respondent's age is approximately 43 years.

Finally, in turn, the multinomial logistic regression analysis, as well as the corresponding fuzzy reasoning model, did not yield any plausible models for

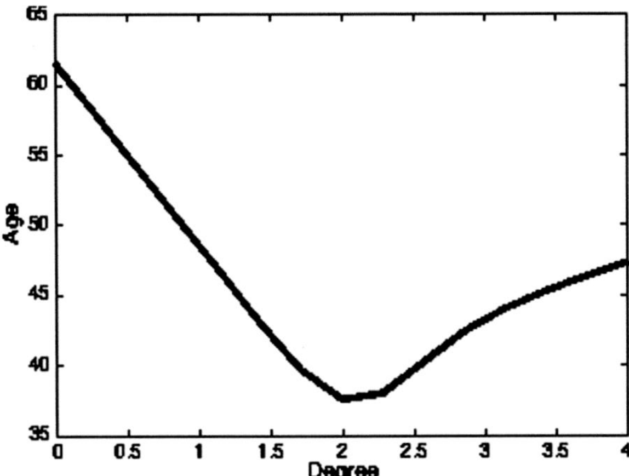

Fig. 23 Group means of age among the degrees of education with with a fuzzy systems of three rules

predicting the voting behavior when age, gender and education were the independent variables.

These statistical analyses and their approximate reasoning counterparts illuminate the potentials for using the concept maps when empiric data sets are examined. In particular, our maps seem promising for pedagogical purposes, because then the students may understand conveniently the interrelationships between the concepts or variables under study. This problem area naturally expects further studies in the future.

6 Conclusions

We applied the concept maps and Zadeh's fuzzy extended logic to model construction. The concept maps enable us to consider such phenomena in which various interrelationships prevail between the concepts or variables. By virtue of the fuzzy extended logic, we may use both linguistic variables and interrelationships in this context as well as computerize them fluently.

We may use our maps for theory formation and model construction when we aim to understand, explain or forecast the phenomena under study. In particular they are usable for pedagogical purposes in the novel learning environments that base on e-learning and intelligent learning paths and materials.

We considered briefly applications on approximate truth valuation as well as on modus ponens and modus tollens syllogisms. These quite simple concept maps may already be used for various reasoning simulations. We also sketched some possibilities for using the concept maps in statistical analysis. All these applications support our assumption that the concept maps with approximate reasoning seem promising methods in particular in education. However, more studies in this area are expected in the future.

Acknowledgments I express my thanks to Prof. Christer Carlsson in Åbo Akademi University who has constantly supported me in my scientific activities. I also thank Academician Janusz Kacprzyk for this opportunity to give my contribution for this book.

References

1. Ausubel, D.: Educational Psychology: A Cognitive View, Holt. Rinehart & Winston, New York (1968)
2. Axelrod, R.: Structure of Decision, The Cognitive Maps of Political Elites. Princeton University Press, Princeton (1976)
3. Bai, S.H., Chen, S.M.: Automatically constructing concept maps based on fuzzy rules for adapting learning systems. Expert Syst. Appl. **35**(1–2), 41–49 (2008)
4. Bandemer, H., Näther, W.: Fuzzy Data Analysis, Kluwer, Dordrecht (1992)
5. Barabasi, A.L.: Linked. Plume, New York (2003)

6. Carnap, R.: The Logical Foundations of Probability, The University of Chicago Press, Chicago (1962)
7. Carvalho, J., Tome, J.: Rule based fuzzy cognitive maps in socio-economic systems. In: Proceedings of the IFSA Congress, pp. 1821–1826. Lisbon (2009)
8. Fullér, R.: Fuzzy Reasoning and Fuzzy Optimization. TUCS General Publications No. 9, Turku Centre for Computer Science, Turku (1998)
9. Gadamer, H.-G.: Truth and Method. Sheed & Ward, London (1975)
10. Haack, S.: Philosophy of Logics. Cambridge University Press, Cambridge (1978)
11. Hempel, C.: Aspects of Scientific Explanation and Other Essays in the Philosophy of Science. The Free Press, New York (1965)
12. Hempel, C.: Philosophy of Natural Science. Prentice Hall, Englewood Cliffs (1966)
13. Hintikka, J., Suppes, P. (eds.): Information and Inference. Reidel, Dordrecht (1970)
14. Kemeny, J., Oppenheim, P.: Degree of factual support. Philos. Sci. **19**(4), 307–324 (1952)
15. Kim, S., Lee, C.: Fuzzy Implications of fuzzy cognitive map with emphasis on fuzzy causal relationship and fuzzy partially causal relationship. Fuzzy Sets Syst. **97**(3), 303–3013 (1998)
16. Kosko, B.: Fuzzy Engineering, Prentice Hall, Upper Saddle River (1997)
17. Lau, R., et al.: Towards fuzzy domain ontology based concept map generation for E-learning. Lecture Notes in Computer Science, vol. 4823, pp. 90–101 (2008)
18. Miao, Y., et al.: Dynamical cognitive network—an extension of fuzzy cognitive map. IEEE Trans. Fuzzy Syst. **9**(5), 760–770 (2001)
19. Niiniluoto, I.: Tieteellinen päättely ja selittäminen (Scientific Reasoning and Explanation). Otava, Keuruu (1983)
20. Niiniluoto, I.: Truthlikeness. Reidel, Dordrecht (1987)
21. Niskanen, V.A.: A Meta-level approach to approximate probability. In: Setchi, R., et al. (eds.) Lecture Notes in Artificial Intelligence, vol. 6279, pp. 116–123. Springer, Heidelberg (2010)
22. Niskanen, V.A.: Application of approximate reasoning to hypothesis verification. J. Intell. Fuzzy Syst. **21**(5), 331–339 (2010)
23. Niskanen, V.A.: Soft Computing Methods in Human Sciences, Studies in Fuzziness and Soft Computing, vol. 134. Springer, Berlin (2004)
24. Novak, J.: Learning, Creating, and Using Knowledge: Concept Maps as Facilitative Tools in Schools and Corporations. Lawrence Erlbaum Associates Inc., Mahwah (1998)
25. Popper, K.: Conjectures and Refutations. Routledge & Kegan Paul, London (1963)
26. Popper, K.: The Logic of Scientific Discovery. Hutchinson, London (1959)
27. Stylios, C., et al.: Modeling complex systems using fuzzy cognitive maps. IEEE Trans. Syst. Man Cybern. Part A **34**(1), 155–162 (2004)
28. Tseng, S., et al.: A new approach for constructing the concept map. Comput. Educ. **49**, 691–707 (2007)
29. Wenstøp, F.: Quantitative Analysis with Linguistic Values. Fuzzy Sets Syst. **4**, 99–115 (1980)
30. von Wright, G.H.: The Logical Foundations of Probability. Blackwell, Oxford (1957)
31. Zadeh, L.: A note on modal logic and possibility theory. Inf. Sci. **279**, 908–913 (2014)
32. Zadeh, L.: From computing with numbers to computing with words—from manipulation of measurements to manipulation of perceptions. IEEE Trans. Circuits Syst. **45**, 105–119 (1999)
33. Zadeh, L.: From search engines to question answering systems? The problems of world knowledge, relevance, deduction and precisiation. In: Sanchez, E. (ed.) Fuzzy Logic and the Semantic Web. Elsevier, Amsterdam (2006)
34. Zadeh, L.: Fuzzy logic and approximate reasoning. Synthese **30**, 407–428 (1975)
35. Zadeh, L.: Fuzzy logic = computing with words. IEEE Trans. Fuzzy Syst. **2**, 103–111 (1996)
36. Zadeh, L.: Toward extended fuzzy logic—a first step. Fuzzy Sets Syst. **160**, 3175–3181 (2009)
37. Zadeh, L.: Toward a theory of fuzzy information granulation and its centrality in human reasoning and fuzzy logic. Fuzzy Sets Syst. **90**(2), 111–127 (1997)
38. Zadeh, L.: Toward a perception-based theory of probabilistic reasoning with imprecise probabilities. J. Stat. Plan. Inference **105**(2), 233–264 (2002)
39. Zar, J.: Biostatistical Analysis. Prentice Hall, Englewood Cliffs (1984)

Part II
Data Analysis, Decision Making and Systems Modeling

Significant Frequent Item Sets Via Pattern Spectrum Filtering

Christian Borgelt and David Picado-Muiño

Abstract Frequent item set mining often suffers from the grave problem that the number of frequent item sets can be huge, even if they are restricted to closed or maximal item sets: in some cases the size of the output can even exceed the size of the transaction database to analyze. In order to overcome this problem, several approaches have been suggested that try to reduce the output by statistical assessments so that only significant frequent item sets (or association rules derived from them) are reported. In this paper we propose a new method along these lines, which combines data randomization with so-called pattern spectrum filtering, as it has been developed for neural spike train analysis. The former serves the purpose to implicitly represent the null hypothesis of independent items, while the latter helps to cope with the multiple testing problem resulting from a statistical evaluation of found patterns.

1 Introduction

Frequent item set mining (see, e.g., [6, 11] for an overview) has been an area of intense research in data mining since the mid 1990s. Up to the early 2000s the main focus was on developing algorithms that can find all frequent, all closed or all maximal item sets as fast as possible. The substantial efforts devoted to this task led to a variety of very sophisticated algorithms, the best-known of which are Apriori [2], Eclat [28, 29], FP-Growth [12–14], and LCM [21–23]. Since the efficiency problem can be considered solved with these algorithms, the focus has shifted since then to the grave problem that the number of found frequent item sets can be huge, even if they are restricted to closed or maximal item sets: in some cases the size of the output

C. Borgelt (✉) · D. Picado-Muiño
European Centre for Soft Computing, Gonzalo Gutiérrez Quirós s/n,
33600 Mieres, Spain
e-mail: christian@borgelt.net

D. Picado-Muiño
e-mail: david.picado@softcomputing.es

M. Collan et al. (eds.), *Fuzzy Technology*, Studies in Fuzziness
and Soft Computing 335, DOI 10.1007/978-3-319-26986-3_4

can even exceed the size of the transaction database to analyze. As a consequence, relevant frequent item sets (or association rules derived from them) can drown in a sea of irrelevant patterns.

In order to overcome this problem, several approaches have been suggested, which fall mainly into two categories: in the first place, it is tried to reduce the output by statistical assessments so that only significant patterns are reported. Such approaches include mining only part of the data and statistically validating the results on a hold-out subset [25] or executing statistical tests directly in the search [26], corrected by Bonferroni [1, 5], Bonferroni-Holm [15], Benjamini-Hochberg [3] or similar methods for multiple testing. A related approach in the spirit of closed item sets are self-sufficient item sets [27]: item sets the support of which is within expectation (under independence assumptions) are removed. A second line in this category consists in randomization approaches (like [9]), which create surrogate data sets that implicitly encode the null hypothesis.

The second category is the selection of so-called *pattern sets*, for example, a (small) pattern set that covers the data well or exhibits little overlap between its member patterns (low redundancy). Such approaches include finding pattern sets with which the data can be compressed well [19, 24] or in which all patterns contribute to partitioning the data [7]. A general framework for this task, which has become known as *constraint based pattern mining*, has been suggested in [8]. Note that in this second category pattern sets are selected, with an emphasis on the interaction between the patterns, while the approaches in the first category rather try to find patterns that are significant individually.

In this paper we propose an approach that falls into the first category and is closest in spirit to [9], mainly because we also use swap randomization to generate surrogate data sets. However, we consider other randomization methods as well, in particular if the transactional data is derived from a table, that is, if the individual items are actually attribute-value pairs. Our method also goes beyond [9] by considering the significance of individual patterns, while [9] only considered the total number of patterns. Finally, we discuss pattern spectrum filtering as a simple, yet effective way to cope with the multiple testing problem.

The remainder of this paper is organized as follows: in Sect. 2 we briefly review frequent item set mining to introduce notation as well as core concepts. In Sect. 3 we discuss randomization or surrogate data generation methods, with which the null hypothesis of independent items is represented implicitly. Section 4 introduces the notion of a pattern spectrum (adapted from [18]) as a way to handle the multiple testing problem that results from the combinatorial explosion of potential patterns. In Sect. 5 we report about experiments that we carried out with several publicly available data sets that are commonly used for benchmarks. Finally, in Sect. 6, we draw conclusions from our discussion.

2 Mining Frequent Item Sets

Formally, frequent item set mining is the following task: we are given a set $B = \{i_1, \dots, i_n\}$ of *items*, called the *item base*, and a database $T = (t_1, \dots, t_m)$ of *transactions*. An item may, for example, represent a product offered by a shop. In this case the item base represents the set of all products offered by, for example, a supermarket or an online shop. The term *item set* refers to any subset of the item base B. Each transaction is an item set and may represent, in the supermarket setting, a set of products that has been bought by a customer. Since several customers may have bought the exact same set of products, the total of all transactions must be represented as a vector (as above) or as a multiset (or bag). Alternatively, each transaction may be enhanced by a *transaction identifier (tid)*. Note that the item base B is usually not given explicitly, but only implicitly as the union of all transactions, that is, $B = \bigcup_{k \in \{1,\dots,m\}} t_k$.

The *cover* $K_T(I) = \{k \in \{1, \dots, m\} \mid I \subseteq t_k\}$ of an item set $I \subseteq B$ indicates the transactions it is contained in. The *support* $s_T(I)$ of I is the number of these transactions and hence $s_T(I) = |K_T(I)|$. Given a user-specified *minimum support* $s_{\min} \in \mathbb{N}$, an item set I is called *frequent* (in T) iff $s_T(I) \geq s_{\min}$. The goal of frequent item set mining is to find all item sets $I \subseteq B$ that are frequent in the database T and thus, in the supermarket setting, to identify all sets of products that are frequently bought together. Note that frequent item set mining may be defined equivalently based on the *(relative) frequency* $\sigma_T(I) = s_T(I)/m$ of an item set I and a corresponding lower bound $\sigma_{\min} \in (0, 1]$.

A typical problem in frequent item set mining is that the number of patterns is often huge and thus the output can easily exceed the size of the transaction database to mine. In order to mitigate this problem, several restrictions of the set of frequent item sets have been suggested. The two most common are *closed* and *maximal* item sets: a frequent item set $I \in \mathcal{F}_T(s_{\min})$ is called

- a *maximal (frequent) item set* iff $\forall J \supset I : s_T(J) < s_{\min}$;
- a *closed (frequent) item set* iff $\forall J \supset I : s_T(J) < s_T(I)$.

In this paper we mainly consider closed item sets, because they not only preserve knowledge of what item sets are frequent, but also allow us to compute the support of non-closed frequent item sets with a simple formula (see, e.g., [6]).

Frequent item set mining usually follows a simple *divide-and-conquer* scheme that can also be seen as a *depth-first search* (essentially only Apriori uses a *breadth-first search*): for a chosen item i, the problem to find all frequent item sets is split into two subproblems: (1) find all frequent item sets containing i and (2) find all frequent item sets *not* containing i. Each subproblem is then further split based on another item j: find all frequent item sets containing (1.1) both i and j, (1.2) i, but not j, (2.1) j, but not i, (2.2) neither i nor j etc.

All subproblems occurring in this recursion can be defined by a *conditional transaction database* and a *prefix*. The prefix is a set of items that has to be added to all frequent item sets that are discovered in the conditional transaction database. Formally, all subproblems are pairs $S = (C, P)$, where C is a conditional database and $P \subseteq B$ is a prefix. The initial problem, with which the recursion is started, is $S = (T, \emptyset)$, where T is the given transaction database.

A subproblem $S_0 = (C_0, P_0)$ is processed as follows: choose an item $i \in B_0$, where B_0 is the set of items occurring in C_0. This choice is, in principle, arbitrary, but often follows some predefined order of the items. If $s_{C_0}(\{i\}) \geq s_{\min}$, then report the item set $P_0 \cup \{i\}$ as frequent with the support $s_{C_0}(\{i\})$, and form the subproblem $S_1 = (C_1, P_1)$ with $P_1 = P_0 \cup \{i\}$. The conditional database C_1 comprises all transactions in C_0 that contain the item i, but with the item i removed. This also implies that transactions that contain no other item than i are entirely removed: no empty transactions are ever kept. If C_1 is not empty, process S_1 recursively. In any case (that is, regardless of whether $s_{C_0}(\{i\}) \geq s_{\min}$ or not), form the subproblem $S_2 = (C_2, P_2)$, where $P_2 = P_0$. The conditional database C_2 comprises *all* transactions in C_0 (including those that do not contain the item i), but again with the item i (and resulting empty transactions) removed. If the database C_2 is not empty, process S_2 recursively.

Concrete algorithms following this scheme differ mainly in how they represent the conditional transaction databases and how they derive a conditional transaction database for a split item from a given database. Details about such algorithms (like Eclat, FP-Growth, or LCM) can be found, for example, in [6, 11].

3 Surrogate Data Generation

The general idea of data randomization or surrogate data generation is to represent the null hypothesis (usually an independence hypothesis; here: independence of the items) not explicitly by a data model, but implicitly by data sets that are generated in such a way that their occurrence probability is (approximately) equal to their occurrence probability under the null hypothesis. Such an approach has the advantage that it needs no explicit data model, which in many cases may be difficult to specify, but can start from the given data. This data is modified in random ways to obtain data that are at least analogous to those that could be sampled under conditions in which the null hypothesis holds.

A randomization or surrogate data approach also makes it usually easier to preserve certain frame conditions and properties of the data to analyze that one may want to keep, in order not to taint the test result by having destroyed features that the data possess, but in which one is not directly interested. In the case of transactional data, such features are the number of items, the number of transactions, the size of the transactions and the (relative) frequency of the items. That is, for standard transactional data, we want a randomization method that only changes the composition of the given transactions, but keeps their sizes and the overall occurrence frequencies of the individual items.

A very simple method satisfying these constraints is *swap randomization* [9], which is best explained with the help of how it modifies a binary matrix representation of a transaction database. In such a representation each column refers to an item, each row to a transaction, and a matrix element is 1 iff the item corresponding to the

Fig. 1 A single swap of swap randomization in a matrix representation

element's column is contained in the transaction corresponding to the element's row. Otherwise the element is 0. Swap randomization consists in executing a large number of *swaps* like the one depicted in Fig. 1. Each swap affects two items and two transactions. Each of the transactions contains one item, but not the other; the swap exchanges the items between the transactions.

In a set representation, as we used it in Sect. 2, a swap can be described as follows: let t_j and t_k be two transactions with $t_j - t_k \neq \emptyset$ and $t_k - t_j \neq \emptyset$, that is, each transaction contains at least one item not contained in the other. Then we choose $i_j \in t_j - t_k$ and $i_k \in t_k - t_j$ and replace t_j and t_k with $t'_j = (t_j - \{i_j\}) \cup \{i_k\}$ and $t'_k = (t_k - \{i_k\}) \cup \{i_j\}$ thus exchanging the items between the transactions. Such a swap has the clear advantage that it obviously maintains the sizes of the transactions as well as the (exact) occurrence frequencies of the items.

If a sufficiently large number of swaps is carried out (in [9] it is recommended to use a number in the order of the 1 s in a binary matrix representation of the data), the resulting transaction database can be seen as being sampled from the null hypothesis of independent items, because all (systematic, non-random) co-occurrences of items have been sufficiently destroyed. Note that it is advisable to apply swap randomization to already generated surrogates to further randomize the data, rather than to start always from the original data. In this way the number of swaps may also be reduced for later surrogates. In our implementation we execute as many swaps as there are 1 s in a binary matrix representation only for the first surrogate, but only half that number for every later surrogate. This provides a good trade-off between speed and independence of the data sets.

An obvious alternative consists in retrieving the (overall) item probability distribution and randomly sampling from it to fill the given transactions with new items (taking care, of course, that no item is sampled more than once for the same transaction). This methods looks simpler (because one need not find transactions first that satisfy the conditions stated above), but has the drawback that it preserves the item frequencies only in expectation. However, this can be corrected (to some degree) by checking the item distribution in a generated surrogate and then adapting the transactions as follows: if there is a transaction (selected randomly) in which an item i occurs that is over-represented relative to the original data, while it lacks an item j that is under-represented, item i is replaced by item j. This procedure is repeated until the item distribution meets, or is at least sufficiently close to the distribution in the original data. In our experiments we found that it was always possible, with fairly little effort in this direction, to meet the actual item frequency distribution exactly.

While these methods work well for actual transactional data, we also have to take care of the fact that many data sets that might be submitted to frequent item set mining (including many common benchmark data sets) are actually derived from tabular data. That is, the items are actually attribute-value pairs, and thus the transactions are sets $t_k = \{A_1 = a_{1k}, \dots, A_n = a_{nk}\}$, where the A_j, $j = 1, \dots, n$, are attributes and a_{jk} is the value that attribute A_j has in the kth transaction, $k = 1, \dots, m$. For such data the methods described above are not applicable, because we have to ensure that each transaction contains exactly one item for each attribute, which is not guaranteed with the above methods.

To randomize such data we use a *column shuffling scheme*. That is, we generate r permutations $\pi_j, j = 1, \dots, r$, of the numbers $\{1, \dots, m\}$ (one permutation for each attribute), where m is the number of transactions. Then we replace each transaction t_k, $k = 1, \dots, m$, with $t'_k = \{A_1 = a_{1\pi_1(k)}, \dots, A_n = a_{n\pi_n(k)}\}$. This guarantees that each transaction contains one item for each attribute. It only shuffles the attribute values, respecting the domains of the attributes.

Other surrogate data generation methods, which are designed for data over an underlying continuous domain (like a time domain), from which the transactions are derived by (time) binning, are discussed in [17]. Unfortunately, they cannot be transferred directly to the transactional setting, because most of them require the possibility to dither/displace items on a continuous (time) scale.

4 Pattern Spectrum Filtering and Pattern Set Reduction

Trying to single out significant patterns proves to be less simple than it may appear at first sight, since one has to cope with the following two problems: in the first place, one has to find a proper statistic that captures how (un)likely it is to observe a certain pattern under the null hypothesis that items occur independently. Secondly, the huge number of potential patterns causes a severe multiple testing problem, which is not easy to overcome with standard methods. In [18] we provided a fairly extensive discussion in the framework of spike train analysis (trying to find patterns of synchronous activity) and concluded that an approach different to evaluating specific patterns with statistics is needed.

As a solution, *pattern spectrum filtering* was proposed in [18, 20] based on the following insight: even if it is highly unlikely that a *specific group* of z items co-occurs s times, it may still be likely that *some group* of z items co-occurs s times, even if items occur independently. The reason is simply that there are so many possible groups of z items (unless the item base B as well as z are tiny) that even though each group has only a tiny probability of co-occurring s times, it may be almost certain that *one of them* co-occurs s times. As a consequence, since there is no *a-priori* reason to prefer certain sets of z items over others (even though a refined analysis, on which we are working, may take individual item frequencies into account), we should not declare a pattern significant if the occurrence of a counterpart (same size z and same or higher support s) can be explained as a chance event under the null hypothesis of independent items.

Hence we pool patterns with the same *pattern signature* $\langle z, c \rangle$, and collect for each signature the (average) number of patterns that we observe in a sufficiently large number of surrogate data sets. This yields what is called a *pattern spectrum* in [18, 20]. Pattern spectrum filtering keeps only such patterns found in the original data for which no counterpart with the same signature (or a signature with the same z, but larger s) was observed in surrogate data, as such a counterpart would show that the pattern can be explained as a chance event.

While in [18, 20] a pattern spectrum is represented as a bar chart with one bar per signature, this is not feasible for the data sets we consider in this paper, due to the usually much larger support values. Rather we depict a pattern spectrum as a bar chart with one bar per pattern size z, the height of which represents the largest support $s_{\max}(z)$ that we observed for patterns of this size in surrogate data sets. An example of such a pattern spectrum is shown in the top part of Fig. 2 (mind the logarithmic scale). Note that this reduced representation, although less rich in information, still contains all that is relevant, namely the support border, below which we discard patterns found in the original data.

Note also that pattern spectrum filtering still suffers from a certain amount of *multiple testing*: every pair $\langle z, c \rangle$ that is found in the original data gives rise to one

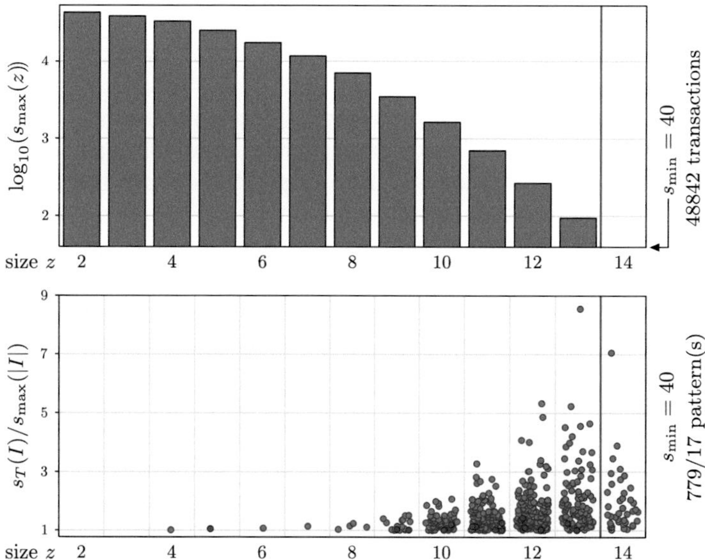

Fig. 2 Pattern spectrum (*top*) and filtered patterns (*bottom*) of the census data. Note the logarithmic scale in the top diagram. The *red line* marks the end of the pattern spectrum: no larger patterns were observed in surrogate data sets. The horizontal position of the *dots* representing the patterns in each size bin of the *bottom* diagram is random (to reduce the dot overlap). Reduced patterns are marked in *red*

test. However, the pairs $\langle z, c\rangle$ are *much fewer* than the number of specific item sets. As a consequence, simple approaches like *Bonferroni correction* [1, 5] become feasible, with which the number of needed surrogate data sets can be computed [18]: given a desired overall significance level α and the number k of pattern signatures to test, at least k/α surrogate data sets have to be analyzed.

As a further filtering step, *pattern set reduction* was proposed in [20] to take care of the fact that an actual pattern induces other, spurious patterns that are subsets, supersets or overlap patterns. These patterns are reduced with the help of a preference relation between patterns and the principle that only patterns are kept to which no other pattern is preferred. Here we adopt the following preference relation: let $X, Y \subseteq B$ be two patterns with $Y \subseteq X$ and let $z_X = |X|$ and $z_Y = |Y|$ be their sizes and s_X and s_y their support values. Finally, let $s_{\max}(z)$ be the largest support of a pattern of size z observed in surrogate data. Then the excess support of Y (relative to X) can be explained as a chance event if $\phi_1 = (s_Y - s_X + 1 \leq s_{\max}(z_Y))$ holds and the excess items in X (relative to Y) can be explained as a chance event if $\phi_2 = (s_X \leq s_{\max}(z_X - z_Y + 2))$ holds. Finally, we use $\phi_3 = ((z_X - 1)s_X \geq (z_Y - 1)s_Y))$ as a heuristic tie-breaker if both ϕ_1 and ϕ_2 hold. As a consequence, the set X is preferred to the set Y iff $\phi_1 \wedge (\neg\phi_2 \vee \phi_3)$ and the set Y is preferred to the set X iff $\phi_2 \wedge (\neg\phi_1 \vee \neg\phi_3)$. Otherwise X and Y are not comparable. More details, especially the reasoning underlying the conditions ϕ_1 and ϕ_2, can be found in [20].

5 Experiments

We implemented the described surrogate data generation methods as well as pattern spectrum filtering in C and made the essential functions available as a Python extension library, which simplifies setting up scripts for the experiments. Pattern set reduction was then implemented on top of this library in Python.

As data sets we chose common benchmark data sets, like the `census`, `chess`, `mushroom`, and `breast` data sets from the UCI machine learning repository [4], the `BMS-Webview-1` data set (or `webview1` for short) from the KDD cup 2000 [16], as well as the `retail`, `accidents` and `kosarak` data sets from the FIMI repository [10]. However, due to reasons of space we can only present some of the results, for which we selected `census`, `breast`, `webview1` and `retail`. The first two of these data sets are actually tabular data, and therefore we applied the column shuffling scheme described above, while the last two are genuinely transactional data, which we processed with swap randomization.

For all data sets we generated and analyzed 10,000 surrogate data sets and ranked the filtered item sets by how far they are from the support border of the pattern spectrum (using the ratio $s_T(I)/s_{\max}(|I|)$, where $s_T(I)$ is the support of I in the transactional database T). A summary of the number of transactions, minimum (absolute) support values, and discovered closed frequent patterns before and after pattern spectrum filtering is shown in Table 1.

Table 1 Data sets for which we present results in this paper together with their sizes, the minimum support used for mining and the number of found patterns

			Closed patterns		
Data set	Trans.	s_{min}	Unfiltered	Filtered	Reduced
census	48842	40	850932	779	17
breast	350	20	965	323	1
webview1	59602	60	3974	259	42
retail	88162	45	19242	3	1

Table 2 Top-ranked closed frequent item sets in the census data

z	s	q	items
12	382	1.425	country=United-States edu_num=10 education=Some-college salary<=50K loss=none gain=none hours=half-time marital=Never-married relationship=Own-child age=young sex=Female workclass=Private
12	362	1.351	country=United-States edu_num=10 education=Some-college salary<=50K loss=none gain=none hours=full-time marital=Never-married relationship=Own-child age=young sex=Male workclass=Private
11	882	1.256	country=United-States edu_num=13 education=Bachelors salary>50K loss=none age=middle-aged marital=Married-civ-spouse relationship=Husband sex=Male race=White workclass=Private

On the census data (see Fig. 2), our filtering methods reduce the huge number of 850932 closed frequent patterns that are found with minimum support $s_{\text{min}} = 40$ to merely 17 statistically significant patterns. The top 3 patterns are shown in Table 2, which are nicely interpretable. The first two capture the children of a family that work directly after finishing college, the third pattern captures upper middle class husbands or family fathers. The differences of the first two patterns, which are highlighted in blue, are interesting to observe (Fig. 3).

On the webview1 data (see Fig. 4) the 3974 closed frequent item sets that are found with minimum support $s_{\text{min}} = 60$ are reduced to merely 42. The top ranked of these patterns are shown in Table 3. Due to the numerical encoding of the items, they are difficult to interpret without any data dictionary, though.

On the retail data (see Fig. 5) the large number of 19242 closed frequent item sets found with minimum support $s_{\text{min}} = 45$ is reduced to the single pattern $I = \{39, 41, 48\}$ with $s_I = 7366$ and $s_I/s_{\text{max}}(3) = 1.12133$. Again an interpretation is difficult, due to the numeric encoding of the items.

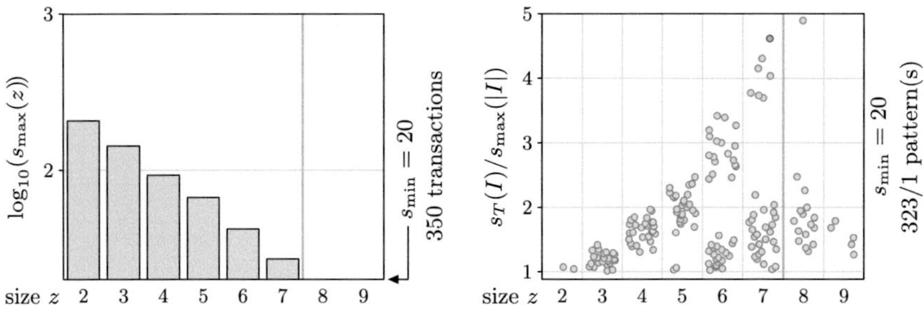

Fig. 3 Pattern spectrum (*left*) and filtered patterns (*right*) of the `breast` data

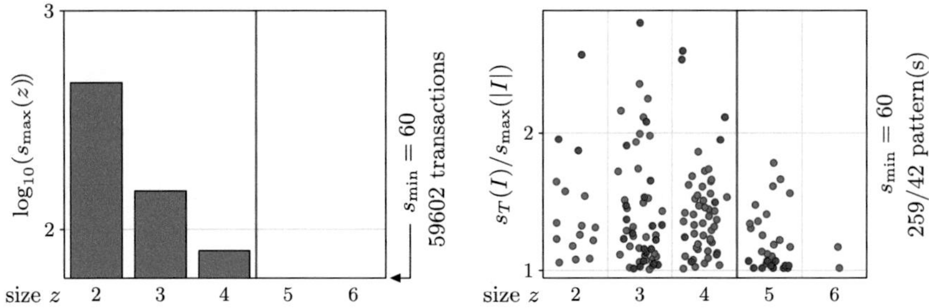

Fig. 4 Pattern spectrum (*left*) and filtered patterns (*right*) of the `webview1` data

Table 3 Top-ranked closed frequent item sets in the `webview1` data

z	s	q	Items
3	417	2.780	10295 10307 10311
4	205	2.562	10295 10307 10311 10315
2	1204	2.561	33449 33469
4	200	2.500	10311 12487 12703 32213

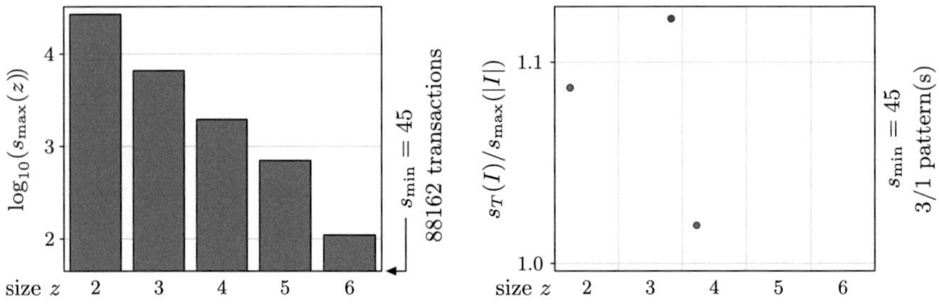

Fig. 5 Pattern spectrum (*left*) and filtered patterns (*right*) of the `retail` data

6 Conclusions and Future Work

We demonstrated how data randomization or surrogate data generation together with pattern spectrum filtering and pattern set reduction can effectively reduce found (closed) frequent item sets to statistically significant ones. The reduction is often tremendous and leaves a user with a manageable number of patterns that is feasible to check manually. A shortcoming of our current method is, however, that it treats all item sets alike, regardless of the frequency of the individual items. We are currently working on an extension that allows for different support borders depending on the expected support of an item set as computed from the individual item frequencies under an independence assumption. Although this is likely to increase the number of filtered patterns, it may enable the method to detect significant item sets consisting of less frequent items.

Software and Source Code

Python and C implementations of the described surrogate data generation and frequent item set filtering procedures can be found at this URL: www.borgelt.net/pyfim.html.

References

1. Abdi, H.: Bonferroni and Šidák corrections for multiple comparisons. In: Salkind, N.J. (ed.) Encyclopedia of Measurement and Statistics, pp. 103–107. Sage Publications, Thousand Oaks (2007)
2. Agrawal, R., Srikant, R.: Fast algorithms for mining association rules. In: Proceedings of the 20th International Conference on Very Large Databases (VLDB 1994, Santiago de Chile), pp. 487–499. Morgan Kaufmann, San Mateo (1994)
3. Benjamini, Y., Hochberg, Y.: Controlling the false discovery rate: a practical and powerful approach to multiple testing. J. R. Stat. Soc. Ser. B (Methodological) **57**(1), 289–300. Blackwell, Oxford, United Kingdom (1995)
4. Blake, C.L., Merz, C.J.: UCI Repository of Machine Learning Databases. Department of Information and Computer Science, University of California at Irvine, CA (1998). http://www.ics.uci.edu/~mlearn/MLRepository.html
5. Bonferroni, C.E.: Il calcolo delle assicurazioni su gruppi di teste. Studi in Onore del Professore Salvatore Ortu Carboni, pp. 13–60. Bardi, Rome (1935)
6. Borgelt, C.: Frequent item set mining. wiley interdisciplinary reviews (WIREs): data mining and knowledge discovery **2**(6), 437–456 (2012). doi:10.1002/widm.1074, Wiley, Chichester, United Kingdom
7. Bringmann, B., Zimmermann, A.: The chosen few: on identifying valuable patterns. In: Proceedings of the 7th IEEE International Conference on Data Mining (ICDM 2007, Omaha, NE), pp. 63–72. IEEE Press, Piscataway, NJ (2007)
8. De Raedt, L., Zimmermann, A: Constraint-based pattern set mining. In: Proceedings of the 7th IEEE International Conference on Data Mining (ICDM 2007, Omaha, NE), pp. 237–248. IEEE Press, Piscataway, NJ (2007)
9. Gionis, A., Mannila, H., Mielikäinen, T., Tsaparas, P.: Assessing data mining results via swap randomization. ACM Trans. Knowl. Discov. Data **1**(3), 14 (2007). ACM Press, New York

10. Goethals, B.: Frequent Itemset Mining Implementations Repository. University of Antwerp, Belgium (2003). http://fimi.ua.ac.be/
11. Goethals, B.: Frequent set mining. Data Mining and Knowledge Discovery Handbook, pp. 321–338. Springer, Berlin (2010)
12. Grahne, G., Zhu, J.: Efficiently using prefix-trees in mining frequent itemsets. In: Proceedings Workshop Frequent Item Set Mining Implementations (FIMI 2003, Melbourne, FL), vol. 90. CEUR Workshop Proceedings, Aachen, Germany (2003)
13. Grahne, G., Zhu, J.: Reducing the main memory consumptions of FPmax* and FPclose. In: Proceedings Workshop Frequent Item Set Mining Implementations (FIMI 2004, Brighton, UK), vol. 126, CEUR Workshop Proceedings, Aachen, Germany (2004)
14. Han, J., Pei, J., Yin, Y.: Mining frequent patterns without candidate generation. In: Proceedings of the19th ACM International Conference on Management of Data (SIGMOD 2000, Dallas, TX), pp. 1–12. ACM Press, New York (2000)
15. Holm, S.: A simple sequentially rejective multiple test procedure. Scand. J. Stat. **6**(2), 65–70 (1979). Wiley, Chichester, United Kingdom
16. Kohavi, R., Bradley, C.E., Frasca, B., Mason, L., Zheng, Z.: KDD-Cup 2000 organizers' report: peeling the onion. SIGKDD Exploration **2**(2), 86–93 (2000) .ACM Press, New York
17. Louis, S., Borgelt, C., Grün, S.: Generation and selection of surrogate methods for correlation analysis. In: Grün, S., Rotter, S. (eds.) Analysis of Parallel Spike Trains, pp. 359–382. Springer, Berlin (2010)
18. Picado-Muiño, D., Borgelt, C., Berger, D., Gerstein, G.L., Grün, S.: Finding neural assemblies with frequent item set mining. Front. Neuroinformatics **7**(9) (2013). doi:10.3389/fninf.2013.00009, Frontiers Media, Lausanne, Switzerland
19. Siebes, A., Vreeken, J., van Leeuwen, M., Item Sets that Compress. In: Proceedings SIAM International Conference on Data Mining (SDM 2006, Bethesda, MD), pp. 393–404. Society for Industrial and Applied Mathematics, Philadelphia, PA, USA (2006)
20. Torre, E., Picado-Muiño, D., Denker, M., Borgelt, C., Grün, S.: Statistical evaluation of synchronous spike patterns extracted by frequent item set mining. Front. Comput. Neurosc. **7**(132) (2013). doi:10.3389/fninf.2013.00132. Frontiers Media, Lausanne, Switzerland
21. Uno, T., Asai, T., Uchida, Y., Arimura, H.: LCM: an efficient algorithm for enumerating frequent closed item sets. In: Proceedings Workshop on Frequent Item Set Mining Implementations (FIMI 2003, Melbourne, FL), vol. 90. CEUR Workshop Proceedings, TU Aachen, Germany (2003)
22. Uno, T., Kiyomi, M., Arimura, H.: LCM ver. 2: efficient mining algorithms for frequent/closed/maximal itemsets. In: Proceedings Workshop Frequent Item Set Mining Implementations (FIMI 2004, Brighton, UK), vol.126. CEUR Workshop Proceedings, Aachen, Germany (2004)
23. Uno, T., Kiyomi, M., Arimura, H.: LCM ver. 3: collaboration of array, bitmap and prefix tree for frequent itemset mining. In: Proceedings of the 1st Open Source Data Mining on Frequent Pattern Mining Implementations (OSDM 2005, Chicago, IL), pp. 77–86. ACM Press, New York, (2005)
24. Vreeken, J., van Leeuwen, M., Siebes, A.: Krimp: mining itemsets that compress. Data Min. Knowl. Discov. **23**(1), 169–214 (2011)
25. Webb, G.I.: Discovering significant patterns. Mach. Learn. **68**(1), 1–33 (2007)
26. Webb, G.I.: Layered critical values: a powerful direct-adjustment approach to discovering significant patterns. Mach. Learn. **71**(2–3), 307–323 (2008)
27. Webb, G.I.: Self-sufficient itemsets: an approach to screening potentially interesting associations between items. ACM Trans. Knowl. Discov. Data (TKDD) **4**(1), 3 (2010)
28. Zaki, M.J., Parthasarathy, S., Ogihara, M., Li, W.: New algorithms for fast discovery of association rules. In: Proceedings of the 3rd International Confernece on Knowledge Discovery and Data Mining (KDD 1997, Newport Beach, CA), pp. 283–296. AAAI Press, Menlo Park, CA, USA (1997)
29. Zaki, M.J., Gouda, K.: Fast vertical mining using diffsets. In: Proceedings of the 9th ACM International Conference on Knowledge Discovery and Data Mining (KDD 2003, Washington, DC), pp. 326–335. ACM Press, New York, NY, USA (2003)

A Novel Genetic Fuzzy System for Regression Problems

Adriano S. Koshiyama, Marley M.B.R. Vellasco and Ricardo Tanscheit

Abstract Solving a regression problem is equivalent to finding a model that relates the behavior of an output or response variable to a given set of input or explanatory variables. An example of such a problem would be that of a company that wishes to evaluate how the demand for its product varies in accordance to its and other competitors' prices. Another example could be the assessment of an increase in electricity consumption due to weather changes. In such problems, it is important to obtain not only accurate predictions but also interpretable models that can tell which features, and their relationship, are the most relevant. In order to meet both requirements—linguistic interpretability and reasonable accuracy—this work presents a novel Genetic Fuzzy System (GFS), called Genetic Programming Fuzzy Inference System for Regression problems (GPFIS-Regress). This GFS makes use of Multi-Gene Genetic Programming to build the premises of fuzzy rules, including in it t-norms, negation and linguistic hedge operators. In a subsequent stage, GPFIS-Regress defines a consequent term that is more compatible with a given premise and makes use of aggregation operators to weigh fuzzy rules in accordance with their influence on the problem. The system has been evaluated on a set of benchmarks and has also been compared to other GFSs, showing competitive results in terms of accuracy and interpretability issues.

1 Introduction

Regression problems are widely reported in the literature [1, 4, 17, 24, 30]. Generalized Linear Models [27], Neural Networks [18] and Genetic Programming [23] tend to provide solutions with high accuracy. However, high precision is not

A.S. Koshiyama (✉) · M.M.B.R. Vellasco · R. Tanscheit
Department of Electrical Engineering, Pontifical Catholic University of Rio de Janeiro, Rio de Janeiro, RJ, Brazil
e-mail: adriano@ele.puc-rio.br

M.M.B.R. Vellasco
e-mail: marley@ele.puc-rio.br

R. Tanscheit
e-mail: ricardo@ele.puc-rio.br

M. Collan et al. (eds.), *Fuzzy Technology*, Studies in Fuzziness and Soft Computing 335, DOI 10.1007/978-3-319-26986-3_5

always associated to a reasonable interpretability, that is, it may be difficult to identify, in linguistic terms, the relation between the response variable (output) and the explanatory variables (inputs).

A GFS integrates a Fuzzy Inference System (FIS) and a Genetic Based Meta-Heuristic (GBMH), which is based on Darwinian concepts of natural selection and genetic recombination. Therefore, a GFS provides fair accuracy and linguistic interpretability (FIS component) through the automatic learning of its parameters and rules (GBMH component) by using information extracted from a dataset or a plant. The number of works related to GFSs applied to regression problems has increased over the years and are mostly based on improving the Genetic Based Meta-Heuristic counterpart of GFSs by using Multi-Objective Evolutionary Algorithms [1, 5, 31]. In general most of these works do not explore linguistic hedges and negation operators. Procedures for the selection of consequent terms have not been reported and few works weigh fuzzy rules. In addition GFSs based on Genetic Programming have never been applied to regression problems.

This work presents a novel GFS called Genetic Programming Fuzzy Inference System for Regression problems (GPFIS-Regress). The main characteristics of this model are: (*i*) it makes use of Multi-Gene Genetic Programming [21, 34], a Genetic Programming generalization that works on a single-objective framework, which can be more reliable computationally in some situations than multi-objective approaches; (*ii*) it employs aggregation, negation and linguistic hedge operators in a simplified manner; (*iii*) it applies some heuristics to define the consequent term best suited to a given antecedent term.

This work is organized as follows: Sect. 2.1 presents some related works on GFSs applied to regression problems and Sect. 2.2. covers the main concepts of the GBMH used in GPFIS-Regress: Multi-Gene Genetic Programming. Section 3 presents the GPFIS-Regress model; case studies are dealt with in Sect. 4. Section 5 concludes the work.

2 Background

2.1 Related Works

In general, GFSs designed for solving regression problems are similar to those devised for classification. This is due to the similarity between those problems, except for the output variable: in regression the consequent term is a fuzzy set, while in classification it is a classical set. Nevertheless, in both cases interpretability is a relevant requirement. Therefore, most works in this subject employ Multi-Objective Evolutionary Algorithms (MOEAs) as the GBMH for rule base synthesis. One of the few that does not follow this concept is that of Alcalá et al. [2], which presents one of the first applications of 2-tuple fuzzy linguistic representation [20]. In this work a GFS, based on a Genetic Algorithm (GA), learns both the granularity and the displacement of the membership functions for each variable. Wang and Mendel's algorithm [35] is used for rule generation. This model is applied to two real cases.

The work of Antonelli et al. [5] proposes a multi-objective GFS to generate a Mamdani-type FIS, with reasonable accuracy and rule base compactness. This system learns the granularity and the fuzzy rule set (a typical knowledge-base discovery approach). It introduces the concept of virtual and concrete rule base: the virtual one is based on the highest number of membership functions for each variable, while the concrete one is based on the values observed in the individual of the MOEA population. This algorithm was applied to two benchmarks for regression.

Pulkkinen and Koivisto [31] present a GFS that learns most of the FIS parameters. A MOEA is employed for fine-tuning membership functions and for defining the granularity and the fuzzy rule base. A feature selection procedure is performed before initialization and an adaptable solution from Wang & Mendel's algorithm is included in an individual as an initial seed. This model performs equally or better than other recent multi-objective and single-objective GFSs for six benchmark problems.

The recent work of Alcalá et al. [1] uses a MOEA for accuracy and comprehension maximization. It defines membership functions granularities for each variable and uses Wang & Mendel's algorithm for rule generation. During the evolutionary process, membership functions are displaced following a 2-tuple fuzzy linguistic representation, as stated earlier. In a post-processing stage fine-tuning of the membership functions is performed. The proposed approach compares favorably to four other GFSs for 17 benchmark datasets.

Benítez and Casillas [8] present a novel multi-objective GFS to deal with high-dimensional problems through a hierarchical structure. This model explores the concept of Fuzzy Inference Subsystems, which compose the hierarchical structure of a unique FIS. The MOEA has a 2-tuple fuzzy linguistic representation that indicates the displacement degree of triangular membership functions and which variables will belong to a subsystem. The fuzzy rule base is learned through Wang & Mendel's algorithm. This approach is compared to other GFSs for five benchmark problems.

Finally, Márquez et al. [26] employ a MOEA to adapt the conjunction operator (a parametric t-norm) that combines the premise terms of each fuzzy rule in order to maximize total accuracy and reduce the number of fuzzy rules. An initial rule base is generated through Wang & Mendel's approach [35], followed by a screening mechanism for rule set reduction. The codification also includes a binary segment that indicates which rules are considered in the system, as well as an integer value that represents the parametric t-norm to be used in a specific rule. An experimental study carried out with 17 datasets of different complexities attests the effectiveness of the mechanism, despite the large number of fuzzy rules.

2.2 Multi-Gene Genetic Programming

Genetic Programming (GP) [23, 30] belongs to the Evolutionary Computation field. Typically, it employs a population of individuals, each of them denoted by

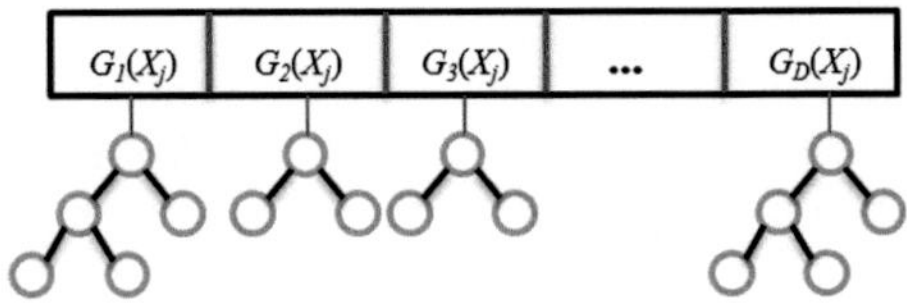

Fig. 1 Example of multi-gene individual

a tree structure that codifies a mathematical equation that describes the relationship between the output Y and a set of input variables X_j $(j = 1, \ldots, J)$. Based on these ideas, Multi-Gene Genetic Programming (MGGP) [15, 17, 21, 34] generalizes GP as it denotes an individual as a structure of trees, also called genes, that similarly receives X_j and tries to predict Y (Fig. 1).

Each individual is composed of D trees or functions $(d = 1, \ldots, D)$ that relate X_j to Y through user-defined mathematical operations. It is easy to verify that MGGP generates solutions similar to those of GP when $D = 1$. In GP terminology, the X_j input variables are included in the Terminal Set, while the mathematical operations (plus, minus, etc.) are part of the Function Set (or Mathematical Operations Set).

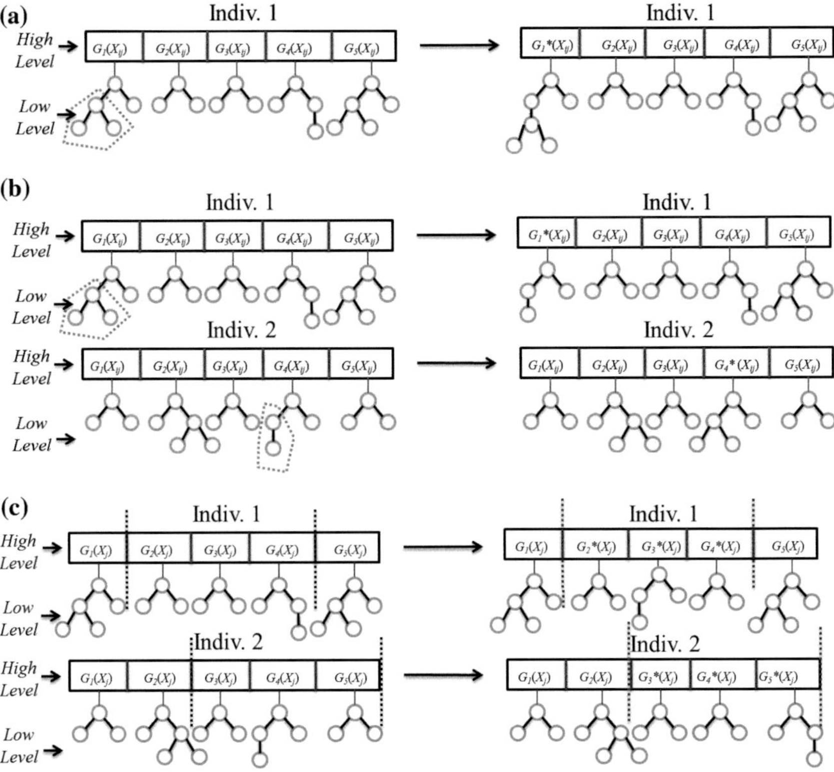

Fig. 2 Application example of MGGP operators: **a** mutation; **b** low level crossover; and **c** high level crossover

With respect to genetic operators, mutation in MGGP is similar to that in GP. As for crossover, the level at which the operation is performed must be specified: it is possible to apply crossover at high and low levels. Figure 2a presents a multi-gene individual with five equations ($D = 5$) accomplishing a mutation, while Fig. 2b shows the low level crossover operation.

The low level is the space where it is possible to manipulate structures (Terminals and Mathematical Operations) of equations present in an individual. In this case, both operations are similar to those performed in GP. The high level, on the other hand, is the space where expressions can be manipulated in a macro way. An example of high level crossover is shown in Fig. 2c. By observing the dashed lines it can be seen that the equations were switched from an individual to the other. The cutting point can be symmetric—the same number of equations is exchanged between individuals—or asymmetric. Intuitively, high level crossover has a deeper effect on the output than low level crossover and mutation have.

In general, the evolutionary process in MGGP differs from that in GP due to the addition of two parameters: maximum number of trees per individual and high level crossover rate. A high value is normally used for the first parameter to assure a smooth evolutionary process. The high level crossover rate, similarly to other genetic operators rates, needs to be adjusted.

3 GPFIS-Regress Model

GPFIS-Regress is a typical Pittsburgh-type GFS [19]. Its development begins with the mapping of crisp values into membership degrees to fuzzy sets (Fuzzification). Then, the fuzzy inference process is divided into three subsections: (*i*) generation of fuzzy rule premises (Formulation); (*ii*) assignment of a consequent term to each premise (Premises Splitting) and (*iii*) aggregation of each activated fuzzy rule (Aggregation). Finally, Defuzzification and Evaluation are performed.

3.1 *Fuzzification*

In regression problems, the main information for predicting the behavior of an an output $y_i \in Y$ $(i = 1, \ldots, n)$ consists of its J attributes or features $x_{ij} \in X_j$ $(j = 1, \ldots, J)$. A total of L fuzzy sets are associated to each jth feature and are given by $A_{lj} = \{(x_{ij}, \mu_{A_{lj}}(x_{ij})) | x_{ij} \in X_j\}$, where $\mu_{A_{lj}} : X_j \rightarrow [0, 1]$ is a membership function that assigns to each observation x_{ij} a membership degree $\mu_{A_{lj}}(x_{ij})$ to a fuzzy set A_{lj}. Similarly, for Y (output variable), K fuzzy sets B_k $(k = 1, \ldots, K)$ are associated.

Three aspects are taken into account when defining membership functions: (*i*) form (triangular, trapezoidal, etc.); (*ii*) support set of $\mu_{A_{lj}}(x_{ij})$; (*iii*) an appropriate linguistic term, qualifying the subspace constituted by $\mu_{A_{lj}}(x_{ij})$ with a context-driven

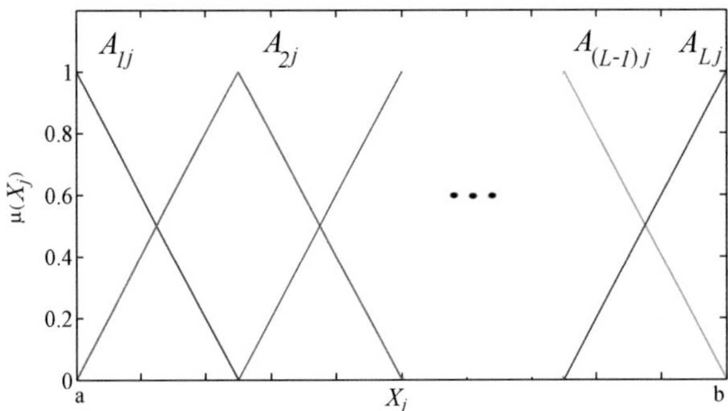

Fig. 3 Membership functions for $x_{ij} \in X_j$ variables. For Y read A_{lj} as B_k

adjective. Ideally, these tasks should be carried out by an expert, whose knowledge would improve comprehensibility. In practice, it is not always easy to find a suitable expert. Therefore it is very common [9, 19, 22] to define membership functions as shown in Fig. 3.

3.2 Fuzzy Inference

3.2.1 Formulation

A fuzzy rule premise is commonly defined by:

"*If* X_1 *is* A_{l1} *and* ... *and* X_j *is* A_{lj} *and* ... *and* X_J *is* X_{lJ}"

or, in mathematical terms:

$$\mu_{A_d}(x_{i1}, \dots, x_{iJ}) = \mu_{A_d}(\mathbf{x}_i) = \mu_{A_{l1}}(x_{i1}) * \dots * \mu_{A_{lJ}}(x_{iJ}) \tag{1}$$

where $\mu_{A_d}(x_{i1}, \dots, x_{iJ}) = \mu_{A_d}(\mathbf{x}_i)$ is the joint membership degree of the ith pattern $\mathbf{x}_i = [x_{i1}, \dots, x_{iJ}]$ with respect to the dth premise ($d = 1, \dots, D$), computed by using a t-norm $*$. A premise can be elaborated by using t-norms, t-conorms, linguistic hedges and negation operators to combine the $\mu_{A_{lj}}(x_{ij})$. As a consequence, the number of possible combinations grows as the number of variables, operators and fuzzy sets increase. Therefore, GPFIS-Regress employs MGGP to search for the most promising combinations, i.e., fuzzy rule premises. Figure 4 exemplifies a typical solution provided by MGGP.

For example, premise 1 represents: $\mu_{A_1}(\mathbf{x}_i) = \mu_{A_{21}}(x_{i1}) * \mu_{A_{32}}(x_{i2})$ and, in linguistic terms, "*If* X_1 *is* A_{21} *and* X_2 *is* A_{32}". Let $\mu_{A_d}(\mathbf{x}_i)$ be the dth premise codified in the dth tree of an MGGP individual. Table 1 presents the components used for reaching the solutions shown in Fig. 4.

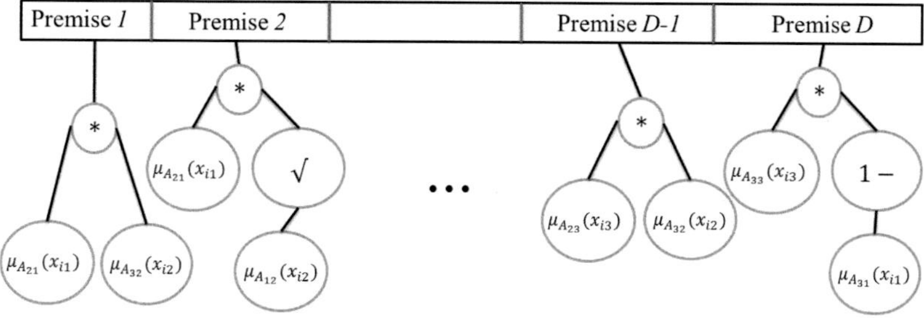

Fig. 4 Example of fuzzy rule premises codified in an MGGP individual

Table 1 Input fuzzy sets and operators to generate solutions in Fig. 4

Input fuzzy sets (Terminals set)	Fuzzy operators set (Functions set)
$\mu_{A_{11}}(y_{t-1}), \ldots, \mu_{A_{L1}}(y_{t-1}), \ldots, \mu_{A_{lp}}(y_{t-p}), \ldots, \mu_{A_{LP}}(y_{t-P})$	t-norm ($*$), linguistic hedge (dilatation operator—$\sqrt{}$) and classical negation operator

In GPFIS-Regress, the set of $\mu_{A_{lj}}(x_{ij})$ represents the Input Fuzzy Sets or, in GP terminology, the Terminal Set, while the Functions Set is replaced by the Fuzzy Operators Set. Thus MGGP is used for obtaining a set of fuzzy rules premises $\mu_{A_d}(\mathbf{x}_i)$. In order to fully develop a fuzzy rule base, it is necessary to define the consequent term best suited to each $\mu_{A_d}(\mathbf{x}_i)$.

3.2.2 Premises Splitting

There are two ways to define which consequent term is best suited to a fuzzy rule premise: (*i*) allow a GBMH to perform this search (a common procedure in several works); or (*ii*) employ methods that directly draw information from the dataset so as to connect a premise to a consequent term. In GPFIS-Regress the second option has been adopted in order to prevent a premise with a large coverage in the dataset, or able to predict a certain region of the output, to be associated to an unsuitable consequent term. Instead of searching for all elements of a fuzzy rule, as a GBMH does, GPFIS-Regress measures the compatibility between $\mu_{A_d}(\mathbf{x}_i)$ and the consequent terms. This also promotes reduction of the search space.

In this sense, the Similarity Degree (SD_k) between the $\mu_{A_d}(\mathbf{x}_i)$ and the consequent terms is employed:

$$SD_k = \min(1 - \frac{\sum_{i=1}^{n} |\mu_{A_d}(\mathbf{x}_i) - \mu_{B_k}(y_i)|}{n}, I_{\{0,1\}}) \in [0, 1] \tag{2}$$

where $\sum_{i=1}^{n} |\mu_{A_d}(\mathbf{x}_i) - \mu_{B_k}(y_i)|$ is the Manhattan distance between the dth premise and the kth consequent term, while $I_{\{0,1\}}$ is an indicator variable, which takes value 0 when $\mu_{A_d}(\mathbf{x}_i) = 0$, $\forall i$, and 1 otherwise. When $\mu_{A_d}(\mathbf{x}_i) = \mu_{B_k}(y_i)$ for all t, then $FCD_k = 1$, i.e., premise and consequent term are totally similar. A consequent term for $\mu_{A_d}(\mathbf{x}_i)$ is selected as the kth consequent which maximize SD_k. A premise with $SD_k = 0$, for all k, is not associated to any consequent term (and not considered as a fuzzy rule).

3.2.3 Aggregation

A premise associated to the kth consequent term (i.e. a fuzzy rule) is denoted by $\mu_{A_{d^{(k)}}}(\mathbf{x}_i)$, which, in linguistic terms, means: "If X_1 is A_{l1}, and ..., and X_J is A_{lJ}, then Y is B_k". Therefore, the whole fuzzy rule base is given by $\mu_{A_{1^{(k)}}}(\mathbf{x}_i),\ldots,\ \mu_{A_{D^{(k)}}}(\mathbf{x}_i)$, $\forall k = 1,\ldots,K$. A new pattern $\mathbf{x}_i^*$ may have a non-zero membership degree to several premises, associated either to the same or to different consequent terms. In order to generate a consensual value, the aggregation step tries to combine the activation degrees of all fuzzy rules associated to the same consequent term.

Consider $D^{(k)}$ as the number of fuzzy rules associated to kth consequent term ($d^{(k)} = 1^{(k)}, 2^{(k)} \ldots, D^{(k)}$). Given an aggregation operator $g : [0,1]^{D^{(k)}} \rightarrow [0,1]$ (see [7, 10]), the predicted membership degree of $\mathbf{x}_i^*$ to each kth consequent term—$\hat{\mu}_{B_k}(y_i^*)$—is computed by:

$$\hat{\mu}_{B_1}(y_i^*) = g[\mu_{A_{1^{(1)}}}(\mathbf{x}_i^*), \ldots, \mu_{A_{D^{(1)}}}(\mathbf{x}_i^*)] \tag{3}$$

$$\hat{\mu}_{B_2}(y_i^*) = g[\mu_{A_{1^{(2)}}}(\mathbf{x}_i^*), \ldots, \mu_{A_{D^{(2)}}}(\mathbf{x}_i^*)] \tag{4}$$

$$\ldots$$

$$\hat{\mu}_{B_K}(y_i^*) = g[\mu_{A_{1^{(K)}}}(\mathbf{x}_i^*), \ldots, \mu_{A_{D^{(K)}}}(\mathbf{x}_i^*)] \tag{5}$$

There are many aggregation operators available (e.g., see [6, 10, 36]), the Maximum being the most widely used [29]. Nevertheless other operators such as arithmetic and weighted averages may also be used. As for weighted arithmetic mean, it is necessary to solve a Restricted Least Squares problem (RLS) in order to establish the weights:

$$min : \sum_{i=1}^{n} (\hat{\mu}_{B_k}(y_i) - \sum_{d^{(k)}=1}^{D^{(k)}} w_{d^{(k)}} \mu_{A_{d^{(k)}}}(\mathbf{x}_i))^2 \tag{6}$$

$$s.t. : \sum_{d^{(k)}=1}^{D^{(k)}} w_{d^{(k)}} = 1 \ and \ w_{d^{(k)}} \geq 0$$

where $w_{d^{(k)}}$ is the weight or the influence degree of $\mu_{A_{d^{(k)}}}(\mathbf{x}_i)$ in the prediction of elements related to the kth consequent term. This is a typical Quadratic Programming problem, the solution of which is easily computed by using algorithms discussed in [11, 33]. This aggregation procedure is called Weighted Average by Restricted Least Squares (WARLS).

3.3 Defuzzification

Proposition 1 *Consider* $y_i \in Y$, *with* $a \leq y_i \leq b$ *where* $a, b \in \mathbb{R}$, *and, associated to* Y, K *triangular membership functions, normal, 2-overlapped*[1] *and strongly partitioned (identical to Fig. 3). Then* y_i *can be rewritten as:*

$$y_i = c_1 \mu_{B_1}(y_i) + c_2 \mu_{B_2}(y_i) + \cdots + c_J \mu_{B_K}(y_i) \tag{7}$$

where $c_1, \ldots, c_K$ *is the "center"—*$\mu_{B_k}(c_k) = 1$*—of each kth membership function.*

The proof can be found in in [28]. This linear combination, which is a defuzzification procedure, is usually known as the Height Method. From this proposition, the following conclusions can be drawn:

1. If $\mu_{B_k}(y_i)$ is known, then y_i is also known.
2. If only a prediction $\hat{\mu}_{B_k}(y_i)$ of $\mu_{B_k}(y_i)$ is known, such that $\sup_{y_t} |\mu_{B_k}(y_i) - \hat{\mu}_{B_k}(y_i)| \leq \varepsilon$, when $\varepsilon \to 0$ the defuzzification output $\hat{y}_i$ that approximates y_i is given by:

$$\hat{y}_i = c_1 \hat{\mu}_{B_1}(y_i) + c_2 \hat{\mu}_{B_2}(y_i) + \cdots + c_K \hat{\mu}_{B_K}(y_t) \tag{8}$$

When $\hat{\mu}_{B_k}(y_i) \approx \mu_{B_k}(y_i)$ is not verified, the Mean of Maximum or the Center of Gravity [32] defuzzification methods may lead to a better performance. However, due to the widespread use of strongly partitioned fuzzy sets in the experiments with GPFIS-Regress, a normalized version of the Height Method (8) has been employed:

$$\hat{y}_i = \frac{c_1 \hat{\mu}_{B_1}(y_i) + \cdots + c_K \hat{\mu}_{B_K}(y_i)}{\hat{\mu}_{B_1}(y_i) + \cdots + \hat{\mu}_{B_K}(y_i)} \tag{9}$$

It is now possible to evaluate an individual of GPFIS-Regress by using $\hat{y}_i$.

3.4 Evaluation

The Evaluation procedure in GPFIS-Regress is defined by a primary objective—error minimization—and a secondary objective—complexity reduction. The primary objective is responsible for ranking individuals in the population, while the secondary one is used as a tiebreaker criteria.

A simple fitness function for regression problems is the Mean Squared Error (*MSE*):

$$MSE = \frac{\sum_{i=1}^{n} (y_i - \hat{y}_i)^2}{2 * n} \tag{10}$$

---

[1] A fuzzy set is normal if it has some element with maximum membership equal to 1. Also, fuzzy sets are 2-overlapped if $\min(\mu_{B_u}(y_i), \mu_{B_z}(y_i), \mu_{B_v}(y_i)) = 0, \forall u, v, z \in k = 1, \ldots, K$.

The best individual in the population is the solution which minimizes (10). GPFIS-Regress tries to reduce the complexity of the rule base by employing a simple heuristic: Lexicographic Parsimony Pressure [25]. This technique is only used in the selection phase: given two individuals with the same fitness, the best one is that with fewer nodes. Fewer nodes indicate rules with fewer antecedent elements, linguistic hedges and negation operators, as well as few premises ($\mu_{A_d}(\mathbf{x}_i)$), and, therefore, a small fuzzy rule set. After evaluation, a set of individuals is selected (through a tournament procedure) and recombined. This process is repeated until a stopping criteria is met. When this occurs, the final population is returned.

4 Case Studies

4.1 Experiments Description

Among the SFGs designed for solving regression problems, the Fast and Scalable Multi-Objective Genetic Fuzzy System (FS-MOGFS) [1] has been used in the experiments. In contrast to other works [2, 5, 8, 26, 31], FS-MOGFS has been chosen because:

1. it makes use of 17 regression datasets, where five of them are highly scalable and high-dimensional;
2. it presents a comparison between three different GFSs;
3. it describes in detail the parameters used for each model and the number of evaluations performed. Furthermore, the results show accuracy (training and test sets) and rule base compactness (average number of rules and of antecedents elements per rule).

In its basic version, FS-MOGFS consists of:

- Each chromosome (C) has two parts ($C = C_1 \cup C_2$): C_1 represents the number of triangular and uniformly distributed membership functions and $C_2 = [\alpha_1, \alpha_2, \ldots, \alpha_J]$, where each α_j is a degree of displacement of the jth variable [2]. To obtain the best possible values for C, the model incorporates a Multi-Objective Genetic Algorithm (MOGA) based on SPEA2 [1]. The two objectives are: minimize the Mean Squared Error and the number of rules.
- In order to build the complete knowledge base (rules and membership functions), rule extraction via Wang & Mendel's algorithm is performed for each chromosome [35]. The Mamdani-type SIF employs the minimum for t-norm and implication, and center of gravity for defuzzification.

Extensions of FS-MOGFS have resulted in two other models: (*i*) **FS-MOGFS**^{e}– identical to FS-MOGFS, but with fast error computation by leaving aside a portion of the database; (*ii*) **FS-MOGFS+TUN**: similar to the previous one, but with fine tuning of membership functions parameters [16]. This model provided the best results

Table 2 Databases considered in experiments

Database	Acronym	J	n	Database	Acronym	J	n
Electrical maintenance	ELE	4	1056	Mortgage	MOR	15	1049
Auto MPG6	MPG6	5	398	Treasury	TRE	15	1049
Auto MPG8	MPG8	7	398	Baseball	BAS	16	337
Analcat	ANA	7	4052	MV artificial domain	MV	10	40768
Abalone	ABA	8	4177	Elevators	ELV	18	16559
Stock	STP	9	950	Computer-activity	CA	21	8192
Weather izmir	WIZ	9	1461	Ailerons	AIL	40	13750
Weather ankara	WAN	9	1609	The insurance company	TIC	85	9822
Forest fires	FOR	12	51				

and was therefore used for comparison with GPFIS-Regress. Databases shown in Table 2 [1] have been considered in case studies.

Five of the 17 databases are of high dimensionality: ELV, AIL, MV, CA e TIC; they have been obtained from the KEEL repository [1]. Similarly to the procedure adopted in Alcalá et al. [1], 100,000 evaluations (population size = 100 and number of generations = 1000) have been carried out in each execution. The remaining parameters are shown in Table 3. With six repeats of 5-fold cross-validation, GPFIS-Regress was executed 30 times. The metrics shown for each database are the average for the 30 trained models. The Mean Squared Error has been used as the fitness function [1].

It should be noted that preliminary tests considered three, five and seven fuzzy sets. As the results did not show any relevant difference as far as accuracy was concerned, five strongly partitioned fuzzy sets (Fig. 3) have been used throughout the experiments, as stated in Table 3.

In addition to FS-MOGFS+TUN, three other SFGs were used for comparison:

- GR-MF [13]: employs an evolutionary algorithm to define granularity and membership functions parameters of a Mamdani-type SIF. The Wang & Mendel method [35] is used for rule generation.

Table 3 GPFIS-Regress main configuration

Parameter	Value
Population size	100
Number of generations	1000
Tree maximum depth	5
Tournament size	2
High level crossover rate	50 %
Low level crossover rate	85 %
Mutation rate	15 %
Elitism rate	1 %
Input fuzzy sets	5 fuzzy sets, displayed like Fig. 3
Fuzzy operators	Product, classical negation and square-root
Partitioning method	SD
Aggregation operator	WARLS
Defuzzification	Height method

- GA-WM [12]: a GA is used for synthesizing granularity and the support of triangular membership functions, as well as for defining the universe of discourse. The rule base is also obtained through Wang & Mendel's algorithm.
- GLD-WM [2]: similar to FS-MOGFS+TUN with respect to granularity and membership functions displacement. Wang & Mendel's algorithm is used for rule generation. Final tuning of membership functions is not performed.

Statistical analysis have followed recommendations from [1, 14] and have been performed in the KEEL software [3], with a significance level of 0,1 ($\alpha = 0.10$).

4.2 Results and Discussion

Table 4 shows the results obtained with GPFIS-Regress and their counterpart GFSs for each database in terms of MSE, average number of rules and of antecedent elements per rule. Results for models other than GPFIS-Regress have been taken from [1]. In general GPFIS-Regress has provided better results in 58 % of cases, followed by FS-MOGFS+TUN with 23 %. GLD-WM has performed better for one single database; the remaining SFGs performed below those three. In high-dimensional problems, GPFIS-Regress as attained better results for three of the five databases.

Table 5 presents results for the Friedman test and Holm method for low-dimensional databases, given a significance level of 10 % [1]. As GPFIS-Regress presented the lowest rank (1.5417), it was chosen as the reference model. It can be observed that GPFIS-Regress achieved higher accuracy than GR-MF, GA-WM and

Table 4 Results for GPFIS-Regress and other GFSs in termos of EQM

Data base	GR-MF			GA-WM			GLD-WM			FS-MOGFS+TUN			GPFIS-regress		
	R/A	Training	Test	R/A	Training	Test	R/A	Training	Test	R/A	Training	Test	R/A	Training	Test
ELE	97/4	16645	18637	47/4	17230	18977	33/4	11483	13384	**9/2**	8803	**9842**	16/3	14695	16818
σ		2319	3386		2501	3195		1085	1978		739	1391		815	493
MPG6	243/5	1.423	28.933	186/5	1.879	8.824	82/5	2.294	4.387	22/3	2.778	4.548	**18/3**	2.890	**4.003**
σ		0.073	8.633		0.235	6.079		0.249	0.899		0.220	1.047		0.066	0.336
MPG8	262/7	1.356	49.36	214/7	1.563	15.216	135/7	1.709	4.782	24/3	2.725	4.381	**18/3**	2.780	**4.087**
σ		0.104	16.2		0.183	9.13		0.170	1.445		0.294	0.909		0.091	0.116
ANA	148/7	0.005	0.017	150/7	0.003	0.008	92/7	0.006	0.008	17/3	0.003	**0.003**	**8/3**	0.002	**0.003**
σ		0.001	0.008		0.001	0.005		0.001	0.004		0.000	0.001		0.000	0.000
ABA	498/8	2.358	2.885	143/8	2.433	2.549	31/8	2.487	2.545	**10/3**	2.393	2.454	16/3	2.370	**2.425**
σ		0.052	0.263		0.052	0.163		0.078	0.170		0.092	0.163		0.068	0.154
STP	343/9	0.4	1.543	344/9	0.389	2.192	217/9	0.299	**0.435**	25/3	0.724	0.892	**17/3**	1.143	1.283
σ		0.019	2.484		0.017	3.168		0.025	0.067		0.112	0.154		0.173	0.256
WIZ	331/9	1.176	9.602	218/9	1.233	3.529	107/9	0.926	1.150	**15/3**	0.867	1.011	17/3	0.811	**0.878**
σ		0.077	8.879		0.065	4.023		0.041	0.123		0.040	0.177		0.046	0.040
WAN	397/9	1.406	7.381	279/9	1.522	2.82	133/9	1.111	2.075	**11/2**	1.313	1.581	15/2	1.307	**1.367**
σ		0.067	5.404		0.065	2.825		0.077	1.407		0.174	0.580		0.079	0.092
FOR	396/12	113	3300	395/12	47	3693	377/12	49	3847	33/3	1593	**2406**	**8/3**	1476	2446
σ		17	2207		24	2787		18	2714		570	2161		868	2456
MOR	209/15	0.03	0.176	160/15	0.02	0.093	78/15	0.016	0.022	**9/3**	0.015	0.018	15/3	0.013	**0.015**
σ		0.002	0.28		0.003	0.147		0.002	0.005		0.004	0.012		0.001	0.002

(continued)

Table 4 (continued)

Data base	GR-MF			GA-WM			GLD-WM			FS-MOGFS+TUN			GPFIS-regress		
TRE	189/15	0.066	0.144	136/15	0.045	0.064	70/15	0.033	0.045	**11/3**	0.030	0.040	15/3	0.031	**0.037**
σ		0.011	0.191		0.007	0.046		0.005	0.015		0.004	0.012		0.002	0.002
BAS	262/16	0.255	12.439	262/16	0.202	11.706	244/16	0.138	3.610	21/6	1.305	**2.699**	**16/3**	1.469	3.037
		0.02	2.177		0.031	2.562		0.014	0.621		0.172	0.620		0.067	0.246
MV	–	–	–	–	–	–	–	–	–	16/3	0,159	**0,160**	**10/3**	2,607	2,599
σ	–	–	–	–	–	–	–	–	–		0,031	0,032		1,499	1,490
ELV	–	–	–	–	–	–	–	–	–	**8/3**	0,900	0,900	12/3	0,875	**0,886**
σ	–	–	–	–	–	–	–	–	–		0,200	0,200		0,106	0,108
CA	–	–	–	–	–	–	–	–	–	15/5	4,763	5,063	**15/4**	4,885	**5,060**
σ	–	–	–	–	–	–	–	–	–		0,404	0,760		0,729	0,739
AIL	–	–	–	–	–	–	–	–	–	20/4	1,864	1,905	**17/3**	1,829	**1,858**
σ	–	–	–	–	–	–	–	–	–		0,221	0,233		0,003	0,003
TIC	–	–	–	–	–	–	–	–	–	25/7	0,026	**0,027**	**11/4**	0,026	**0,027**
σ	–	–	–	–	–	–	–	–	–		0,000	0,002		0,000	0,001

σ = standard deviation of EQM. Results for Training/Test must be multiplied by 10^5, 10^{-5} and 10^{-8} for BAS, ELV and AIL respectively. R/A—Averages of Rules and Antecedent Elements per Rule

Table 5 Results for Friedman test and Holm method

i	Model	Rank	
4	GR-MF	4.6667	
3	GA-WM	4.1250	
2	GLD-WM	2.8750	
1	FS-MOGFS+TUN	1.7917	
0	GPFIS	**1.5417**	
Test	p-value		
Friedman	<0.0001		
Method	$z = (R_0 - R_i)/SE$	p-value	Holm
GR-MF	4.8412	<0.0001	0.0250
GA-WM	4.0020	<0.0001	0.0333
GLD-WM	2.0655	0.0388	0.0500
FS-MOGFS+TUN	0.3872	0.6985	0.1000

GLD-WM have (p-value < 0.05). This has not been verified for GPFIS-Regress and FS-MOGFS+TUN (p-value > 0.10).

If GPFIS-Regress and FS-MOGFS+TUN are singled out for comparison, it can be observed that the former has achieved better results for 10 of the 17 databases, with two ties. The signal test has shown that the differences in results were not significant ($S = 10$, p-value $= 0.3018$). This may be due to the ties and to the small number of databases considered. As for rule base complexity, it can be noted that GPFIS-Regress obtained the most compact one in 53 % of cases.

As far as interpretability and implementation are concerned, GPFIS-Regress has an advantage over FS-MOGFS+TUN in aspects such as: (*i*) makes no change to membership functions parameters; (*ii*) employs a MHG with a single objective, while FS-MOGFS+TUN does a multi-objective search.

5 Conclusion

This work has presented a novel Genetic Fuzzy System for solving regression problems, called GPFIS-Regress, which makes use of Multi-Gene Genetic Programming and a novel way to formulate the Fuzzy Reasoning Method (Formulation-Splitting-Aggregation). GPFIS-Regress has been compared to four other Genetic Fuzzy Systems for 17 datasets of low and high dimensionality. Results have shown the potentialities of the proposed approach with respect to the state-of-art in the Genetic Fuzzy Systems area.

Further developments and experiments shall include: (*i*) evaluation of other t-norm, negation and linguistic hedges operators, as well as the use of t-conorms in rules premises; (*ii*) new premises splitting methods (through other similarity measures) and application of the Restricted Least Squares procedure with some adaptation to associate a more suitable consequent term to a given premise; (*iii*) evaluation of other aggregation operators, such as nonlinear ones (weighted geometric mean, etc.); this may provide better results mostly in terms of accuracy. A fine-tuning of membership functions and Genetic Programming set-up parameters shall also be considered.

References

1. Alcalá, R., Gacto, M.J., Herrera, F.: A fast and scalable multiobjective genetic fuzzy system for linguistic fuzzy modeling in high-dimensional regression problems. Fuzzy Syst. IEEE Trans. **19**(4), 666–681 (2011)
2. Alcalá, R., Alcalá-Fdez, J., Herrera, F., Otero, J.: Genetic learning of accurate and compact fuzzy rule based systems based on the 2-tuples linguistic representation. Int. J. Approx. Reason. **44**(1), 45–64 (2007)
3. Alcalá-Fdez, J., Fernandez, A., Luengo, J., Derrac, J., García, S., Sánchez, L., Herrera, F.: Keel data-mining software tool: data set repository, integration of algorithms and experimental analysis framework. J. Mult.-Valued Logic Soft Comput. **17**(2–3), 255–287 (2011)
4. Angelov, P., Buswell, R.: Identification of evolving fuzzy rule-based models. Fuzzy Syst. IEEE Trans. **10**(5), 667–677 (2002)
5. Antonelli, M., Ducange, P., Lazzerini, B., Marcelloni, F.: Learning concurrently partition granularities and rule bases of mamdani fuzzy systems in a multi-objective evolutionary framework. Int. J. Approx. Reason. **50**(7), 1066–1080 (2009)
6. Beliakov, G., Warren, J.: Appropriate choice of aggregation operators in fuzzy decision support systems. IEEE Trans. Fuzzy Syst. **9**(6), 773–784 (2001)
7. Beliakov, G., Pradera, A., Calvo, T.: Aggregation Functions: A Guide for Practitioners. Springer Publishing Company, Heidelberg (2008)
8. Benítez, A.D., Casillas, J.: Multi-objective genetic learning of serial hierarchical fuzzy systems for large-scale problems. Soft Comput. **17**(1), 165–194 (2013)
9. Berlanga, F.J., Rivera, A.J., del Jesus, M.J., Herrera, F.: Gp-coach: genetic programming-based learning of compact and accurate fuzzy rule-based classification systems for high-dimensional problems. Inf. Sci. **180**(8), 1183–1200 (2010)
10. Calvo, T., Kolesárová, A., Komorníková, M., Mesiar, R.: Aggregation operators: properties, classes and construction methods. In: Calvo, T., Mayor, G., Mesiar, R. (eds.) Aggregation Operators, Studies in Fuzziness and Soft Computing, vol. 97, pp. 3–104. Physica-Verlag HD (2002)
11. Coleman, T.F., Li, Y.: A reflective newton method for minimizing a quadratic function subject to bounds on some of the variables. SIAM J. Optim. **6**(4), 1040–1058 (1996)
12. Cordón, O., Herrera, F., Magdalena, L., Villar, P.: A genetic learning process for the scaling factors, granularity and contexts of the fuzzy rule-based system data base. Inf. Sci. **136**(1–4), 85–107 (2001)
13. Cordón, O., Herrera, F., Villar, P.: Generating the knowledge base of a fuzzy rule-based system by the genetic learning of the data base. IEEE Trans. Fuzzy Syst. **9**(4), 667–674 (2001)
14. Derrac, J., García, S., Molina, D., Herrera, F.: A practical tutorial on the use of nonparametric statistical tests as a methodology for comparing evolutionary and swarm intelligence algorithms. Swarm Evol. Comput. **1**(1), 3–18 (2011)
15. Fattah, K.A.: A new approach calculate oil-gas ratio for gas condensate and volatile oil reservoirs using genetic programming. Oil Gas Bus. **1**, 311–323 (2012)
16. Gacto, M.J., Alcalá, R., Herrera, F.: Adaptation and application of multi-objective evolutionary algorithms for rule reduction and parameter tuning of fuzzy rule-based systems. Soft Comput. **13**(5), 419–436 (2008)

17. Gandomi, A.H., Alavi, A.H.: A new multi-gene genetic programming approach to nonlinear system modeling. Neural Comput. Appl. **21**(1), 171–187 (2012)
18. Haykin, S.: Neural Netw. Learn. Mach. Prentice-Hall, New York (2009)
19. Herrera, F.: Genetic fuzzy systems: taxonomy, current research trends and prospects. Evol. Intell. **1**(1), 27–46 (2008)
20. Herrera, F., Martinez, L.: A 2-tuple fuzzy linguistic representation model for computing with words. IEEE Trans. Fuzzy Syst. **8**(6), 746–752 (2000)
21. Hinchliffe, M., Hiden, H., McKay, B., Willis, M., Tham, M., Barton, G.: Modelling chemical process systems using a multi-gene. In: Late Breaking Papers at the Genetic Programming, pp. 56–65, Stanford University, Stanford, June 1996
22. Ishibuchi, H., Yamane, M., Nojima, Y.: Rule weight update in parallel distributed fuzzy genetics-based machine learning with data rotation. In: In IEEE International Conference on Fuzzy Systems, 2013. FUZZ-IEEE 2013, pp. 1–8. IEEE (2013)
23. Koza, J.R.: Genetic Programming: On the Programming of Computers by Means of Natural Selection. MIT Press, Massachusetts (1992)
24. Kutner, M.H., Nachtsheim, C.J., Neter, J., Li, W.: Applied Linear Statistical Models, 8th edn. McGraw-Hill, New York (2005)
25. Luke, S., Panait, L.: Lexicographic parsimony pressure. In: Langdon, W.B., Cantú-Paz, E., Mathias, K., Roy, R., Davis, D., Poli, R., Balakrishnan, K., Honavar, V., Rudolph, G., Wegener, J., Bull, L., Potter, M.A., Schultz, A.C., Miller, J.F., Burke, E., Jonoska, N. (eds) GECCO 2002: Proceedings of the Genetic and Evolutionary Computation Conference, pp. 829–836. Morgan Kaufmann Publishers, New York (2002)
26. Márquez, A.A., Márquez, F.A., Roldán, A.M., Peregrín, A.: An efficient adaptive fuzzy inference system for complex and high dimensional regression problems in linguistic fuzzy modelling. Knowl.-Based Syst. **54**, 42–52 (2013)
27. McCullagh, P., Nelder, J.A.: Generalized Linear Models. Chapman Hall, London (1989)
28. Pedrycz, W.: Granular Computing: Analysis and Design of Intelligent Systems. CRC Press, Boca Raton (2013)
29. Pedrycz, W., Gomide, F.: An Introduction to Fuzzy Sets: Analysis and Design. MIT Press, Massachussets (1998)
30. Poli, R., Langdon, W.B., McPhee, N.F.: A Field Guide to Genetic Programming. Lulu.com, Rayleigh (2008)
31. Pulkkinen, P., Koivisto, H.: A dynamically constrained multiobjective genetic fuzzy system for regression problems. IEEE Trans. Fuzzy Syst. **18**(1), 161–177 (2010)
32. Roychowdhury, S., Pedrycz, W.: A survey of defuzzification strategies. Int. J. Intell. Syst. **16**(6), 679–695 (2001)
33. Schölkopf, B., Smola, A.J.: Learning with Kernels: Support Vector Machines, Regularization, Optimization, and Beyond. MIT Press, Massachussets (2001)
34. Searson, D., Willis, M., Montague, G.: Coevolution of nonlinear pls model components. J. Chemom. **21**(12), 592–603 (2007)
35. Wang, L.X., Mendel, J.M.: Generating fuzzy rules by learning from examples. IEEE Trans. Syst. Man Cybern. **22**(6), 1414–1427 (1992)
36. Yager, R.R., Kacprzyk, J.: The Ordered Weighted Averaging Operators: Theory and Applications. Kluwer, Norwell (1997)

A Multidistance Approach to Consensus Modeling

Silvia Bortot, Mario Fedrizzi, Michele Fedrizzi and Ricardo Alberto Marques Pereira

Abstract We investigate the relationship between the soft measure of collective dissensus introduced in (Fedrizzi et al. Int J Intell Syst 14:63–77, 1999; Fedrizzi et al. New Math Nat Comput 3:219–237, 2007; Preferences and Decisions: Models and Applications, Springer, Heidelberg, 2010) and the multidistance approach to consensus evaluation described in (Brunelli et al. IPMU 2012, Part I, CCIS, Springer, Berlin, 2012). The novelty of the contribution consists in the introduction of a particular type of sum-based multidistance used as a measure of dissensus, closely related with the one introduced in (Fedrizzi et al. New Math Nat Comput 3:219–237, 2007). This multidistance is characterized by the application of a subadditive filtering function whose effect is that of emphasizing small distances and attenuating large ones. An illustrative example is then developed in order to compare the new dissensus measure with the OWA-based multidistance obtained assuming that the weights are linearly decreasing with respect to increasing distance values.

Keywords Multidistances · OWA aggregation · Dissensus measures

1 Introduction

Distance-based consensus starting from individual ordinal preference relations was studied first in [25], assuming that the distance measures the sum for all the individuals of the number of pairs of alternatives for which the relative position is different in the individual's and in the group's ordinal preferences. The conditions under which the various methods yield the same consensus ranking have been discussed in [11], in combination with the presentation of the related mathematical programming formulations. Then, an overview of distance minimizing methods introducing a way

S. Bortot · R.A. Marques Pereira
Department of Economics and Management, University of Trento,
Via Inama 5, 38122 Trento, Italy

M. Fedrizzi(✉) · M. Fedrizzi
Department of Industrial Engineering, University of Trento,
Via Sommarive 9, 38123 Trento, Italy
e-mail: mario.fedrizzi@unitn.it

M. Collan et al. (eds.), *Fuzzy Technology*, Studies in Fuzziness and Soft Computing 335, DOI 10.1007/978-3-319-26986-3_6

of measuring the degree of disagreement prevailing in the profile has been introduced in [30]. Under the same theoretical framework, in [10] a further extension was proposed generating a consensual collective order by solving a goal programming problem. Other approaches to distance-based consensus reaching have been developed assuming that individuals are expected to modify their opinions in order to increase the consensus level using mediation process such as Delphi (see [26]). Since during this kind of consensus processes a significant amount of time and resources is used in order to move the individuals' opinions towards a shared group opinion, the problem of minimization of costs becomes relevant. The minimum cost consensus has been addressed in [1] assuming that individuals with a linear cost of changing their opinions are involved, and then extended in [2] for finding the group opinion that minimizes a quadratic cost function.

In the classical social choice-based approach, the notion of consensus has been usually understood in terms of strict and unanimous agreement. However, since decision makers typically have different and conflicting opinions to a lesser or greater extent, the traditional strict meaning of consensus is often unrealistic. The human perception of consensus is typically 'softer', and people are generally willing to accept that consensus has been reached when most actors agree on the preferences associated to the most relevant alternatives.

Combining the fuzzy notion of consensus with the expressive power of linguistic quantifiers, the so-called soft consensus measure in the context of fuzzy preference relations has been introduced in [22, 23].

The soft consensus paradigm proposed in [22] was then reformulated in [14–16]. The linguistic quantifiers in the original soft consensus measure were substituted by smooth scaling functions with an analogous role, and a dynamical model was obtained from the gradient descent optimization of a soft consensus cost function, combining a soft measure of collective dissensus with an individual mechanism of opinion changing aversion. The resulting soft consensus dynamics acts on the network of single preference structures by a combination of a collective process of diffusion and an individual mechanism of inertia.

Introduced as an extension of the crisp model of consensus dynamics described in [14], the fuzzy soft consensus model in [15] substitutes the standard crisp preferences by fuzzy triangular preferences. The fuzzy extension of the soft consensus model is based on the use of a distance measure between triangular fuzzy numbers. In analogy with the standard crisp model, the fuzzy dynamics of preference change towards consensus derives from the gradient descent optimization of the new cost function of the fuzzy soft consensus model. Comprehensive reviews can be found in [8, 17, 18, 20, 21, 24].

In the applications related to group decision making, the ordered weighted averaging (OWA) operator as introduced in [34, 35] has been extensively experienced and then extended to the modeling of consensus. In the approach adopted in [3] a consensus degree is computed for each alternative, under the assumption of alternative independency on each expert. The novelty of the proposed procedure consists in the direct computation of "soft" linguistic degrees of consensus based on a topological approach [13].

An OWA-based consensus operator under the 2-tuple fuzzy linguistic representation model is proposed in [12], and it's shown that it provides an alternative consensus model for group decision making preserving the original preference information and supporting consensus reaching process without moderator. In [19] some linguistic OWA operators are presented to compute consensus measures under unbalanced linguistic preferences, assuming that the consensus model is based on two types of consensus measures, consensus degree and proximity measures. On the basis of OWA aggregation the authors in [32] introduce a type of ordered weighted distance (OWD), whose main characteristic is to relieve or intensify the influence of deviations on the aggregation results by an appropriate assignment of weights. In [33] the OWD measures are used to model a consensus reaching process with linguistic, interval, triangular or trapezoidal fuzzy preference information.

It's well known that the classical notion of distance has been extended to that of multidistance in [27, 28] in order to axiomatically measure how separated the members of a collection of more than two elements are [27, 28]. Consequently, the definition of multidistance aims at measuring the distance for more than two entities and can be used to evaluate the dissensus of preferences expressed by a group of decision makers [9]. A further extension has been introduced in [7] establishing a relationship between multidistances and m-ary adjacency relations through the use of OWA operators. Starting from the results obtained in [4–6], in [7] some connections between valued m-ary relations and multidistances were highlighted leading to the conclusion that the two approaches to measure how m entities of a collection are separated are mutually supportive. Accordingly, it has been shown how m-ary adjacency relations can be modeled on the basis of OWA-based multidistances, and some consensus related optimization problems on m-ary adjacency relations are equivalent to corresponding multidistance minimization problems.

In this paper, a multidistance dissensus measure is introduced as an extension of the relationship between the soft measure of collective dissensus firstly proposed in [14] and the consensus model developed in [7] and based on multidistances. This measure is based on a pairwise distance defined through a subadditive function whose effect is that of emphasizing small distances and attenuating large distances.

The remainder of the paper is organized as follows. In Sect. 2, after introducing the definition of multidistance and a subclass characterized by the sum of pairwise distances, the so-called OWA-based multidistances are defined. In Sect. 3, a particular type of sum-based multidistance is proposed as characterized by the application of a filtering function and it's shown how it can be used as a measure of dissensus closely related with the one introduced in [14] . In Sect. 4 an illustrative example is developed where the new dissensus measure is compared with the OWA-based multidistance where the weighting vector has linearly decreasing entries. The example demonstrates how the choice of this vector is hierarchically propagated in the aggregation process starting from the 2-argument distances until the n-argument multidistances. As a consequence, the resulting dissensus measure takes gradually more into account similar preferences and gives less importance to very discording preferences. In Sect. 5 some concluding remarks are presented.

2 Multidistances as Dissensus Measures

In this section we briefly review the notion of Multidistance and its use in constructing measures of disagreement among a group of decision makers. Let us first describe the problem we are considering. Given n decision makers, let $x_j \in [0, 1]$ represent the preference degree elicited by decision maker j, with $j = 1, \dots, n$, for instance in comparing two possible alternatives. The degree of disagreement between decision maker i and decision maker j is given by the usual distance $d(x_i, x_j) = |x_i - x_j| \in [0, 1]$. In this paper we use multidistances in order to extend this notion and measure the overall disagreement among n decision makers.

Multidistances were introduced by Martin and Mayor in [27, 28]. An important class of multidistances, i.e. functionally expressible multidistances, are studied in [29, 31]. Applications of multidistances in the problem of consensus measuring can be found in [7, 9].

The definition of 'multidistance' aims at extending the usual notion of distance to the case of more than two points. As the distance between two points measure 'how separated' two points of a space are, analogously a multidistance aims to measure 'how separated' the members of a collection of more than two elements are. The definition given in [28] is as follows.

Given a domain $X \subseteq \mathbb{R}$, a multidistance is a function

$$D : \bigcup_{n\geq 1} X^n \to [0, \infty[$$

with the following properties, for all $n = 1, 2, \dots$ and $x_1, \dots, x_n \in X$:
(P1) $D(x_1, \dots, x_n) = 0$ if and only if $x_i = x_j$ for all $i, j = 1, \dots, n$
(P2) $D(x_1, \dots, x_n) = D(x_{\pi(1)}, \dots, x_{\pi(n)})$ for any permutation π of $1, \dots, n$
(P3) $D(x_1, \dots, x_n) \leq D(x_1, y) + \dots + D(x_n, y)$ for all $y \in X$.

Note that (P1), (P2) and (P3) extend the usual three distance axioms. In particular, (P2) refers to symmetry and (P3) extends the triangle inequality.

There are several methods to construct multidistances, each one leading to different properties [7, 9, 28, 29]. In order to evaluate the disagreement among decision makers in a group, in what follows we focus on two methods that we consider particularly interesting and suitable for our framework.

2.1 *Sum-Based Multidistances*

The first method aimed at evaluating the disagreement among the decision makers is based on a multidistance belonging to the class of the 'sum-based multidistances'. As suggested in [28], given a usual definition of distance $d(x_i, x_j)$, a multidistance may be defined on the basis of the sum of the pairwise distances, by multiplying this sum by a sufficiently small value $\lambda(n)$ depending on n. This type of multidistance is called 'sum-based multidistance' and the following result holds:

Given a domain $X \subseteq \mathbb{R}$, the function $D_\lambda : \bigcup_{n\geq 1} X^n \to [0, \infty[$ defined by

$$\begin{cases} D_\lambda(x_1) = 0 \\ D_\lambda(x_1, \dots, x_n) = \lambda(n) \sum_{i,j=1}^{n} d(x_i, x_j) \ , \ n \geq 2 \end{cases} \tag{1}$$

is a multidistance if and only if

(i) $\lambda(2) = \frac{1}{2}$

(ii) $0 < \lambda(n) \leq \frac{1}{2(n-1)}$ for $n \geq 3$.

In Sect. 3 we define a particular type of sum-based multidistance which is characterized by the application of a filtering function f playing a crucial role in emulating a previously introduced dissensus measure [14, 15, 18]. More precisely, we first define a pairwise distance d_f by applying a suitable function f to the usual distance $d(x, y) = |x - y| \in [0, 1]$, for $x, y \in X = [0, 1]$. Then, we define a multidistance D_f by averaging pairwise distances d_f. This sum-based multidistance can be used as a measure of dissensus $D_f : [0, 1]^n \to [0, 1]$ which is closely related with the dissensus measure introduced in [14, 15, 18].

In the next subsection we briefly describe another type of multidistances and define the second method we propose to evaluate the disagreement among the decision makers.

2.2 OWA-Based Multidistances

Another relevant class of multidistances, the so-called 'OWA-based multidistances' [28], is based on OWA aggregation functions [34].

We recall that OWA (ordered weighted averaging) functions were introduced by Yager [34] and form a class of flexible averaging functions. They are defined as follows. An OWA function A is a mapping $A : X^n \to X$, with associated a weighting vector $\mathbf{w} = (w_1, \dots, w_n)$, such that

$$A_\mathbf{w}(x_1, \dots, x_n) = \sum_{j=1}^{n} w_j x_{(j)}$$

where $x_{(1)} \leq \cdots \leq x_{(n)} \in X$ and $w_j \geq 0$ for $j = 1, \dots, n$ and $\sum_{j=1}^{n} w_j = 1$.

Martin and Mayor [28] used OWA functions in order to define a particular class of multidistances. The underlying idea is a sort of hierarchical aggregation. First, a 'ternary' multidistance $D(x_1, x_2, x_3)$ is defined by averaging the usual three 'binary' distances $D(x_1, x_2)$, $D(x_1, x_3)$, $D(x_2, x_3)$ using an OWA function $A_\mathbf{w}$ with a 3-dimensional weighting vector $\mathbf{w}$,

$$D(x_1, x_2, x_3) = A_{\mathbf{w}}(D(x_1, x_2), D(x_1, x_3), D(x_2, x_3)).$$

Analogously, multidistance $D(x_1, x_2, x_3, x_4)$ is defined by averaging the four multidistances $D(x_1, x_2, x_3)$, $D(x_1, x_2, x_4)$, $D(x_1, x_3, x_4)$, $D(x_2, x_3, x_4)$ using an OWA function $A_{\mathbf{w}}$ with a 4-dimensional weighting vector **w**,

$$D(x_1, x_2, x_3, x_4) = A_{\mathbf{w}}(D(x_1, x_2, x_3), D(x_1, x_2, x_4), D(x_1, x_3, x_4), D(x_2, x_3, x_4)).$$

Multidistances with n-argument $D(x_1, \ldots, x_n)$ are defined similarly, by means of $(n-1)$-argument multidistances. More formally, the general definition of OWA-based multidistance is as follows. An OWA-based multidistance D is a multiargument function $D : \bigcup_{n \geq 1} X^n \to [0, \infty[$ such that

$$\begin{cases} D(x_1) = 0 \\ D(x_1, x_2) = d(x_1, x_2) \\ D(x_1, \ldots, x_n) = A_{\mathbf{w}}(D(\mathbf{a}_1), \ldots, D(\mathbf{a}_n)), \text{ for } n \geq 3, \end{cases} \tag{2}$$

where $d(x_1, x_2)$ is a usual distance, $\mathbf{w} = (w_1, \ldots, w_n)$ is a weighting vector with $w_j \geq 0$ for $j = 1, \ldots, n$ and $\sum_{j=1}^{n} w_j = 1$, and $\mathbf{a}_j$ is the $(n-1)$-dimensional vector obtained from $(x_1, \ldots, x_n)$ by removing the jth component.

The properties of an OWA-based multidistance are clearly induced by the weighting vector $\mathbf{w} = (w_1, \ldots, w_n)$. We propose to use a vector with linearly decreasing entries,

$$\mathbf{w} = (w_1, \ldots, w_n) = \frac{1}{s_n}(n, n-1, \ldots, 3, 2, 1) \tag{3}$$

where $s_n = \sum_{j=1}^{n} j = \frac{n(n+1)}{2}$. The choice of **w** as in (3) defines a multidistance which is comparable with the multidistance D_f defined in Sect. 2.1 and is also closely related with the dissensus measure introduced in [14, 15, 18]. In Sect. 4 we propose an illustrative example where we numerically compare the multidistance D_f with the OWA-based multidistance (2) with **w** given by (3). The effect of the choice of vector **w** as in (3) is motivated by the goal of emphasizing small multidistance values and simultaneously giving less relevance to large multidistance values. This effect is hierarchically propagated in the aggregation process starting from usual 2-argument distances $D(x_i, x_j)$ until the n-argument multidistance $D(x_1, \ldots, x_n)$. As a consequence, the obtained multidistance (2) with **w** given by (3) defines a measure of dissensus which takes gradually more into account similar preferences and gives less importance to very discordant preferences. Interestingly, this measure is closely related to the dissensus measure defined in the framework of the so called 'soft consensus model' [14, 15, 18] where this effect was induced by a suitable scaling function. Note that this type of dissensus measure is able to detect whether there is a good agreement at least between some decision makers, even if some other decision makers strongly disagree.

3 A Multidistance Approach to Soft Consensus

Consider the domain $X = [0, 1]$ equipped with the usual distance $d(x, y) = |x - y| \in [0, 1]$, for $x, y \in [0, 1]$, with the usual triangular inequalities $|x + y| \leq |x| + |y|$ and $d(x, y) \leq d(x, z) + d(y, z)$, for all $x, y, z \in [0, 1]$.

Consider now the function $f : [0, 1] \to [0, 1]$ defined as

$$f(u) = \frac{2}{\alpha} \ln \left(\frac{1 + e^{\alpha\beta}}{1 + e^{-\alpha(u-\beta)}} \right) \qquad \alpha \neq 0 \tag{4}$$

and $f(u) = u$ for $\alpha = 0$, for $u \in [0, 1]$. The two parameters are $\alpha \in [0, \infty)$ and $\beta \in [0, 1/2]$. Notice that $f(0) = 0$ for any choice of the parameters α, β and

$$f(1) = \frac{2}{\alpha} \ln \left(\frac{1 + e^{\alpha\beta}}{1 + e^{-\alpha(1-\beta)}} \right). \tag{5}$$

We obtain $f(1) = 1$ when $\beta = 1/2$ for any choice of the parameter α. In Fig. 1 we plot the function $f(u)$ with various choices for the parameters α and β. In each plot the diagonal line is associated with $\alpha = 0$.

The function f is strictly increasing and concave for any choice of the parameter, it is strictly concave for $\alpha \in (0, \infty)$. These properties follow straightforwardly from the first and second derivatives of f.

The function f is subadditive, in the sense that $f(u + v) \leq f(u) + f(v)$. The proof is as follows: assuming $u,\ v \in [0, 1]$ and $u + v \neq 0$, concavity of f implies

$$f(u) \geq \frac{v}{u+v} f(0) + \frac{u}{u+v} f(u+v) = \frac{u}{u+v} f(u+v) \tag{6}$$

$$f(v) \geq \frac{u}{u+v} f(0) + \frac{v}{u+v} f(u+v) = \frac{v}{u+v} f(u+v) \tag{7}$$

and therefore we obtain $f(u) + f(v) \geq f(u + v)$ for $u,\ v \in [0, 1]$.

Given that f is subadditive, the composition of the distance d and the function f yields a new distance function denoted $d_f(x, y) = f(d(x, y))$, with the triangle inequality $d_f(x, y) \leq d_f(x, z) + d_f(y, z)$. This triangle inequality is due to the subadditivity of f and is obtained as follows,

$$d(x, y) \leq d(x, z) + d(y, z) \tag{8}$$

$$f(d(x, y)) \leq f(d(x, z) + d(y, z)) \leq f(d(x, z)) + f(d(y, z)) \tag{9}$$

where the first inequality is due to the increasingness of f and the second inequality is due to the subadditivity of f. Finally, we obtain

$$d_f(x, y) \leq d_f(x, z) + d_f(y, z). \tag{10}$$

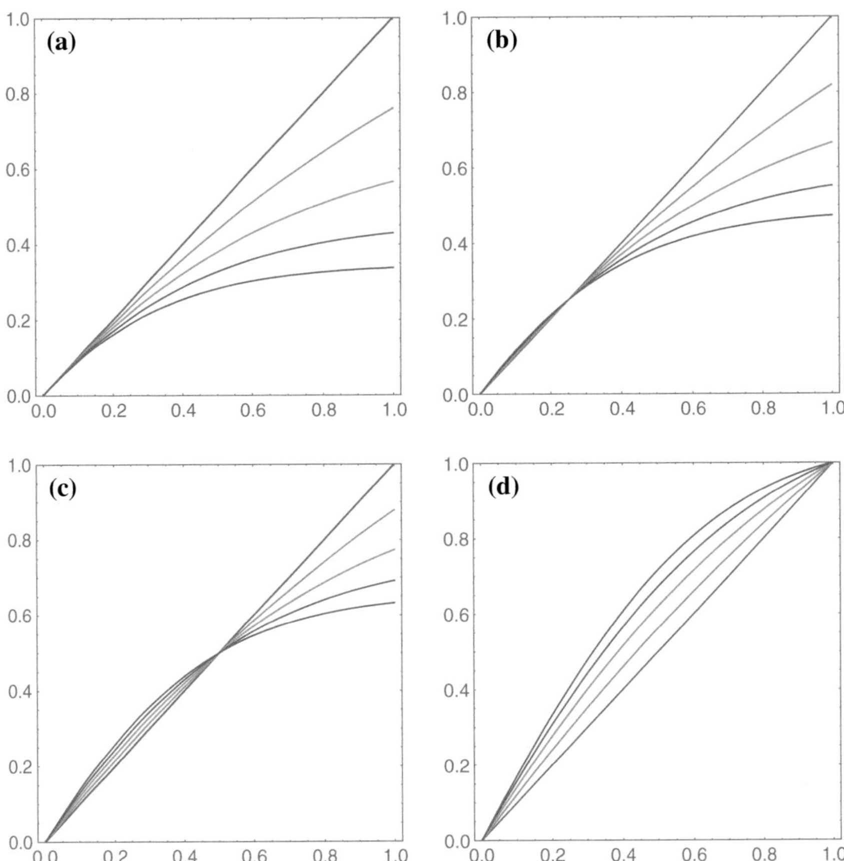

Fig. 1 The function $f(u)$ for $u \in [0,1]$. Each plot is associated with a choice of the parameter β and depicts the graph of $f(u)$ for various choices of the parameter α. **a** $f(u)$ with $\alpha = 0, 1, 2, 4$ and $\beta = 0$. **b** $f(u)$ with $\alpha = 0, 1, 2, 4$ and $\beta = 0.125$. **c** $f(u)$ with $\alpha = 0, 1, 2, 4$ and $\beta = 0.25$. **d** $f(u)$ with $\alpha = 0, 1, 2, 4$ and $\beta = 0.5$

We consider the construction of multidistances based on the distance d_f, in particular by averaging pairwise distances d_f. In this way we build a measure of dissensus $D_f : [0,1]^n \to [0,1]$ which is closely related with the dissensus measure in the soft consensus model. There is essentially a single difference: the basic pairwise distance $d_f(x, y)$ involves $|x - y|$ and not $(x - y)^2$, because the latter is not a distance function. In this way we can obtain a multidistance dissensus measure.

4 An Illustrative Example

We now illustrate with an example the way in which the multidistances constructed by means of the function f can emulate the multidistance construction by means of aggregation with an OWA function A for each dimensionality level.

Assume $n = 4$ and $x_1 = 0.2$, $x_2 = 0.4$, $x_3 = 0.6$, $x_4 = 0.8$. The multidistance constructed by means of the function f is given by

$$D_f(\boldsymbol{x}) = \frac{1}{n(n-1)} \sum_{i,j=1}^{n} d_f(x_i, x_j) \tag{11}$$

$$= \frac{1}{6}\Big(d_f(x_1,x_2) + d_f(x_1,x_3) + d_f(x_1,x_4) + d_f(x_2,x_3) + d_f(x_2,x_4) + d_f(x_3,x_4)\Big)$$

$$= \frac{1}{6}\Big(f(d(x_1,x_2)) + f(d(x_1,x_3)) + f(d(x_1,x_4)) + f(d(x_2,x_3)) + f(d(x_2,x_4)) + f(d(x_3,x_4))\Big)$$

$$= \frac{1}{6}\Big(f(|x_1 - x_2|) + f(|x_1 - x_3|) + f(|x_1 - x_4|) + f(|x_2 - x_3|) + f(|x_2 - x_4|) + f(|x_3 - x_4|)\Big).$$

On the other hand, the multidistance construction by means of aggregation with an OWA function A is given by

$$D_A(\boldsymbol{x}) = D_A(x_1, x_2, x_3, x_4) \tag{12}$$

$$= A\big(D_A(x_1,x_2,x_3), D_A(x_1,x_2,x_4), D_A(x_1,x_3,x_4), D_A(x_2,x_3,x_4)\big)$$

$$= A\Big(A\big(d(x_1,x_2), d(x_1,x_3), d(x_2,x_3)\big), A\big(d(x_1,x_2), d(x_1,x_4), d(x_2,x_4)\big),$$

$$A\big(d(x_1,x_3), d(x_1,x_4), d(x_3,x_4)\big), A\big(d(x_2,x_3), d(x_2,x_4), d(x_3,x_4)\big)\Big)$$

$$= \frac{4}{10}\Big(\frac{3}{6}|x_1 - x_2| + \frac{2}{6}|x_2 - x_3| + \frac{1}{6}|x_1 - x_3|\Big)$$

$$+ \frac{3}{10}\Big(\frac{3}{6}|x_2 - x_3| + \frac{2}{6}|x_3 - x_4| + \frac{1}{6}|x_2 - x_4|\Big)$$

$$+ \frac{2}{10}\Big(\frac{3}{6}|x_1 - x_2| + \frac{2}{6}|x_2 - x_4| + \frac{1}{6}|x_1 - x_4|\Big)$$

$$+ \frac{1}{10}\Big(\frac{3}{6}|x_3 - x_4| + \frac{2}{6}|x_1 - x_3| + \frac{1}{6}|x_1 - x_4|\Big)$$

where the n-dimensional weighting vectors of the OWA functions are as in (3).

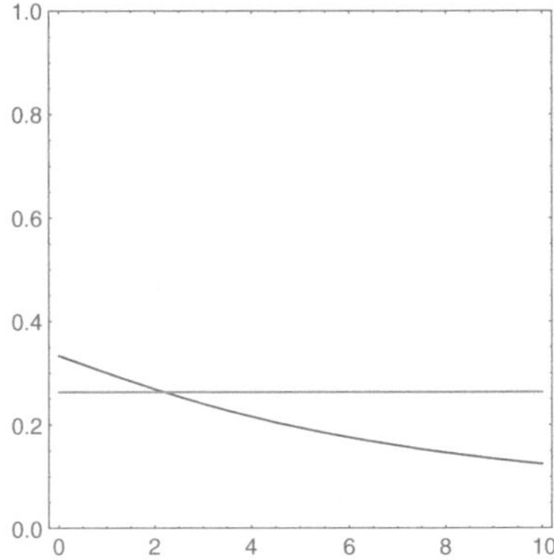

Fig. 2 The multidistance $D_f(\boldsymbol{x})$ as a function of the parameter α with $\beta = 0$, against the multidistance $D_A(\boldsymbol{x}) = 0.26\bar{3}$

The value $D_f(\boldsymbol{x})$ depends on the parameters α, β whereas $D_A(\boldsymbol{x}) = 0.26\bar{3}$. In Fig. 2 we plot $D_f(\boldsymbol{x})$ as a function of the parameter α with $\beta = 0$ against the constant value $D_A(\boldsymbol{x}) = 0.26\bar{3}$ and we see that $D_f(\boldsymbol{x})$ and $D_A(\boldsymbol{x})$ coincide for $\alpha \approx 2.19335$.

5 Conclusions

We introduce a multidistance measure of dissensus within a group of decision makers. The multidistance dissensus measure is based on a fundamental pairwise distance d_f associated with a subadditive function f over the domain $X = [0, 1]$, $d_f(x, y) = f(|x - y|)$. This subadditive function has the effect of emphasizing small distances and attenuating large distances, in analogy with the scaling function which plays a central role in the soft consensus model [14, 15, 18]. Finally, in the context of an illustrative example, the multidistance dissensus measure based on d_f is compared with the dissensus measure obtained through hierarchical aggregation by means of an OWA function with linearly decreasing weights with respect to increasing distance values.

References

1. Ben-Arieh, D., Easton, T.: Multi-criteria group consensus under linear cost opinion elasticity. Decis. Support Syst. **43**(3), 713–721 (2007)
2. Ben-Arieh, D., Easton, T., Evans, B.: Minimum cost consensus with quadratic cost functions. IEEE Trans. Syst. Man Cybern. Part A Syst. Hum. **39**(1), 210–217 (2007)
3. Bordogna, G., Fedrizzi, M., Pasi, G.: A linguistic modeling of consensus in group decision making based on OWA operators. IEEE Trans. Syst. Man Cybern. Part A Syst. Hum. **27**(1), 126–133 (1997)
4. Brunelli, M., Fedrizzi, M.: A fuzzy approach to social network analysis In: Proceedings of ASONAM 2009, pp. 225–230. IEEE Computer Society (2009) doi:10.1109/ASONAM. 2009.72

5. Brunelli, M., Fedrizzi, M., Fedrizzi, M.: OWA-based fuzzy *m*-ary adjacency relations in Social Network Analysis. In: Yager, R.R., Kacprzyk, J., Beliakov, G. (eds.) Recent Developments in the Ordered Weighted Averaging Operators: Theory and Practice. Studies in Fuzziness and Soft Computing, vol. 265, pp. 255–267. Springer, Berlin (2011)
6. Brunelli, M., Fedrizzi, M., Fedrizzi, M.: Fuzzy *m*-ary adjacency relations in social network analysis: optimization and consensus evaluation. Inf. Fusion **17**, 36–45 (2014)
7. Brunelli, M., Fedrizzi, M., Fedrizzi, M., Molinari, F.: On some connections between multidistances and valued m-ary adjacency relations. In: Greco, S., et al. (eds.) IPMU 2012, Part I, CCIS, vol. 297, pp. 201–207. Springer, Berlin (2012)
8. Cabrerizo, F.J., Moreno, J.M., Perez, I.J., Herrera-Viedma, E.: Analyzing consensus approaches in fuzzy group decision making: advantages and drawbacks. Soft Comput. **14**(5), 451–463 (2010)
9. Calvo, T., Martin, J., Mayor, G.: Measures of Disagreement and Aggregation of Preferences based on Multidistances. In: Greco, S., et al. (eds.) IPMU 2012, Part IV, CCIS, vol. 300, pp. 549–558. Springer, Berlin (2012)
10. Contreras, I.: A distance-based consensus model with flexible choice of rank-position weights. Group Decis. Negot. **19**(5), 441–456 (2010)
11. Cook, W.D.: Distance-based and ad hoc consensus models in ordinal preference ranking. Eur. J. Oper. Res. **172**(2), 369–385 (2006)
12. Dong, Y., Xu, Y., Li, H., Feng, B.: The OWA-based consensus operatot under linguistic representation models using position indexes. Eur. J. Oper. Res. **203**(2), 455–463 (2010)
13. Fedrizzi, M., Kacprzyk, J., Nurmi, H.: Consensus degrees under fuzzy majorities and fuzzy preferences using OWA (ordered weighed average) operators. Control Cybern. **22**(4), 77–86 (1993)
14. Fedrizzi, M., Fedrizzi, M., Marques Pereira, R.A.: Soft consensus and network dynamics in group decision making. Int. J. Intell. Syst. **14**, 63–77 (1999)
15. Fedrizzi, M., Fedrizzi, M., Marques Pereira, R.A.: Consensus modelling in group decision making: a dynamical approach based on fuzzy preferences. New Math. Nat. Comput. **3**, 219–237 (2007)
16. Fedrizzi, M., Fedrizzi, M., Marques Pereira, R.A., Brunelli, M.: Consensual dynamics in group decision making with triangular fuzzy numbers. In: Proceedings of the 41st Annual Hawaii International Conference on System Sciences, pp. 70–78. Los Alamitos (2008)
17. Fedrizzi, M., Pasi, G.: Fuzzy logic approaches to consensus modelling in group decision making. Intell. Decis. Policy Mak. Support Syst. **117**, 19–37 (2008)
18. Fedrizzi, M., Fedrizzi, M., Marques Pereira, R.A., Brunelli, M.: The dynamics of consensus in group decision making: investigating the pairwise interactions between fuzzy preferences. In: Greco, S., Marques Pereira, R.A., Squillante, M., Yager, R.R., Kacprzyk, J. (eds.) Preferences and Decisions: Models and Applications, pp. 159–182. Springer, Heidelberg (2010)
19. Herrera-Viedma, E., Cabrerizo, F.J., Perez, I.J., Cobo, M.J., Alonso, S., Herrera, F.: Applying linguistic OWA operators in consensus models under unbalanced linguistic information. In: Yager, R.R., Kacprzyk, J., Beliakov, G. (eds.) Recent Developments in the OWA Operators. Studies in Fuzziness and Soft Computing, vol. 265, pp. 167–186. Springer, Berlin (2011)
20. Herrera-Viedma, E., Garca-Lapresta, J.L., Kacprzyk, J., Fedrizzi, M., Nurmi, H., Zadrożny, S. (eds.): Consensual Processes. Studies in Fuzziness and Soft Computing, vol. 267. Springer, Berlin (2011)
21. Herrera-Viedma, E., Cabrerizo, F., Kacprzyk, J., Pedrycz, W.: A review of soft consensus models in a fuzzy environment. Inf. Fusion **17**, 4–13 (2014)
22. Kacprzyk, J., Fedrizzi, M.: "Soft" consensus measures for monitoring real consensus reaching processes under fuzzy preferences. Control Cybern. **15**(3–4), 309–323 (1986)
23. Kacprzyk, J., Fedrizzi, M.: A "soft" measure of consensus in the setting of partial (fuzzy) preferences. Eur. J. Oper. Res. **34**(3), 316–325 (1988)
24. Kacprzyk, J., Nurmi, H., Fedrizzi, M. (eds.): Consensus Under Fuzziness. International Series in Intelligent Technologies. Kluwer Academic Press, Dordrecht (1997)
25. Kemeny, J.G.: Mathematics without numbers. Daedalus **88**(4), 577–591 (1959)

26. Linstone, H.A., Turoff, M. (eds.): The Delphi Method: Techniques and Applications. Addison Wesley, MA (1975)
27. Martin, J., Mayor, G.: Some Properties of Multi-argument Distances and Fermat Multidistance. In: Hllermeier, E., Kruse, R., Hoffmann, F. (eds.) IPMU 2010, Part I. CCIS, vol. 80, pp. 703–711. Springer, Berlin (2010)
28. Martin, J., Mayor, G.: Multi-argument distances. Fuzzy Sets Syst. **167**, 92–100 (2011)
29. Martin, J., Mayor, G., Valero, O.: Functionally expressible multidistances. In: Proceedings of the 7th conference of the European Society for Fuzzy Logic and Technology (EUSFLAT-2011), pp. 41–46. Atlantis Press, Paris (2011)
30. Meskanen, T., Nurmi, H.: Distance from consensus: a theme and variations. In: Simeone, B., Pukelsheim, F. (eds.) Mathematics and Democracy. Studies in Choice and Welfare, pp. 117–132. Springer, Berlin (2006)
31. Molinari, F.: About a new family of multidistances. Fuzzy Sets and Syst. **195**, 118–122 (2012)
32. Xu, Z., Chen, J.: Ordered weighted distance measure. J.Syst. Sci. Syst. Eng. **17**(4), 432–445 (2008)
33. Xu, Z.: Fuzzy ordered distance measure. Fuzzy Optim. Decis. Mak. **11**(1), 73–87 (2012)
34. Yager, R.R.: On ordered weighted averaging aggregation operators in multi-criteria decision making. IEEE Trans. Syst. Man Cybern. **18**, 183–190 (1988)
35. Yager, R.R.: Applications and extensions of OWA aggregations. Int. J. Man Mach. Stud. **37**(1), 103–122 (1992)

A Consensus Reaching Support System Based on the Concepts of an Ideal and Anti-Ideal Agent and Option

Janusz Kacprzyk, Dominika Gołuńska and Sławomir Zadrożny

Abstract We present an extension of our previous works on a moderator run consensus reaching process in a small group of autonomous decision makers (agents). Our approach is based on fuzzy preferences meant as the testimonies provided by agents, fuzzy majority represented as linguistic quantifiers in Kacprzyk's sense, some fuzzy majority based soft measure of the consensus, proposed by Kacprzyk and Fedrizzi, that is the degree to which: "most of agents agree with their preferences to the most of options", and Kacprzyk and Zadrożny's ideas of a decision support system for consensus reaching, and the use of additional information expressed as linguistic summaries equated with linguistically quantified propositions. Emphasis is on the running of a consensus reaching process by a moderator. To help the moderator run the process in an effective and efficient way, we apply some additional higher-level information, notably in the form of linguistic data summaries as proposed by Kacprzyk and Zadrożny. Here, we extend our approach with a new concept of an ideal and anti-ideal agent and option, though we will concentrate of the case of the ideal and anti-ideal agent as it is more intuitively appealing than that of the ideal and anti-ideal option. The, we use a TOPSIS method based approach that was first outlined in our context by Gołuńska, Kacprzyk and Zadrożny and which boils down to the determination of a solution with the longest distance from the anti-ideal solution and the shortest distance to the ideal one. The improvement of obtaining a higher degree of consensus within the group of agents by using the enhanced moderated consensus reaching support system is illustrated with a numerical example.

J. Kacprzyk (✉) · S. Zadrożny
Polish Academy of Sciences, Systems Research Institute, Ul. Newelska 6, 01-447 Warsaw, Poland
e-mail: kacprzyk@ibspan.waw.pl

S. Zadrożny
e-mail: zadrozny@ibspan.waw.pl

D. Gołuńska
Faculty of Physics and Applied Computer Science, AGH University of Science and Technology, 30-059 Cracow, Poland
e-mail: falkiewicz.d@gmail.com

M. Collan et al. (eds.), *Fuzzy Technology*, Studies in Fuzziness and Soft Computing 335, DOI 10.1007/978-3-319-26986-3_7

1 Introduction

We deal with decision making in a multiperson setting: we have a set of options, i.e. possible choices, and a set of individuals, called here agents, who can play a role of experts, decision makers, … The agents present their testimonies as to some aspect, criterion, … which are here assumed in a general and convenient form of fuzzy preference relations that may adequately and easily represent preferences of a particular agent with respect to a pair of options, represented by some degrees of preference from [0,1]—cf. [2, 3, 8, 11, 13, 14, 26, 27].

Though there are many aspects of group decision making, in this paper we deal with the reaching of a consensus in a group of agents as the first, indispensable (pre) stage of the determination of a group decision which is basically meant as the determination of an option or a set of options that is best accepted by the group. In virtually all cases when the group in question is at some sort of a consensus, then this may usually speed up the reaching of a final group decision and its quality that is usually meant in terms of its acceptance by the group of agents as a whole.

To be more specific, we are concerned with a *consensus reaching process* in a (small) group of (human) autonomous agents (decision makers, experts) who present their testimonies assumed to be in a general form of a fuzzy (graded) preference relation defined in a set of options. We do not consider problems of reaching consensus in society or similar large groups.

Usually, the testimonies of agents significantly differ at the beginning, and the consensus reaching process is run on a step-by-step basis by updating the testimonies of the particular agents, with respect to particular pairs of options, until they become close enough, that is, until the group arrives at some sufficient agreement expressed through a degree of consensus in the sense of Kacprzyk and Fedrizzi [15–17]. Of course, we have to assume the willingness to change testimonies by agents which may involve some psychological apprehension.

The above process is implemented using a specialized group decision support system with an interactive user-friendly interface to view and collect data, share information and opinions between agents and with the moderator, suggest some issues and courses of action, etc. all that meant to facilitate and help effectively and efficiently run the process [28, 29].

In this paper we use such a moderated (moderator run) consensus reaching system the architecture of which was proposed in our previous papers—cf. Gołuńska and Kacprzyk [6], Gołuńska et al. [7], and Gołuńska et al. [8, 9]. Here it is enhanced, first, with some novel concepts and techniques (Kacprzyk and Zadrożny [18–23], Kacprzyk et al. [25]) that are based on the use of some additional information and insight as to the structure of testimonies of agents, and their temporal evolution with respect to changes across the set of agents and options. All those indicators basically point out to those agents, options, and their related aspects which are critical for the effectiveness in the sense of their high impact on a positive change of the degree of consensus.

This is clearly an efficiency-oriented strategy in which only a few of the most promising agents and options are taken into account since this implies the highest effect. Such an approach is commonly employed in almost all works on group decision making and consensus reaching, and will also be assumed here though some new paradigm, based on fairness and equity orientation has also been proposed by Gołuńska and Kacprzyk [6, 7] in which agents and their testimonies with respect to options are treated more fairly, i.e. are taken into account to some extent even if this may not imply the best effect in terms of an increase of the degree of consensus.

In this paper we extend our previous paper (cf. Gołuńska et al. [6]) in which we have employed the idea of the TOPSIS method based on the aggregation of the "closeness to the ideal" and "farness from the anti-ideal" [1, 17]. The latter method was successfully implemented in multicriteria optimization problems [1, 2, 16, 15]. In this paper we employ this technique to find and then effectively and efficiently use the concept of an ideal and anti-ideal agent, and to a lesser extent option, and a TOPSIS method based approach that was first outlined in our context by Gołuńska et al. [6] and which boils down to the determination of a solution with the longest distance from the anti-ideal solution and the shortest distance to the ideal one. The improvement of obtaining a higher degree of consensus within the group of agents by using the enhanced moderated consensus reaching support system is illustrated with a numerical example. We augment this method with more metainformation type additional insights into the very essence of the problem given as linguistic summaries as proposed by Kacprzyk and Zadrożny [22], notably the measures expressing linguistically: for the agents—the response to the omission of an agent and so-called personal consensus degree, and for the options—the response to exclusion and so-called option consensus degree.

In Sect. 2 we present the general scheme of the moderated consensus reaching process considered, in Sect. 3—the basic concepts and properties of the proposed fuzzy-logic-based consensus reaching process, and some additional discussion guiding linguistic summaries, in Sect. 4—the basic definitions of an ideal and anti-deal point as meant in our context, followed by a description of the new approach in Sect. 5 which combines those linguistic summaries mentioned above and meant for the determination of the ideal and anti-ideal agent and option, with elements of TOPSIS [17]. Then, in Sect. 6 we show an example, and in Sect. 7—some concluding remarks and suggestions for future works.

2 The Essence of the Consensus Reaching Process

We basically employ a general framework for supporting consensus reaching in a group of agents due to Kacprzyk et al. [4], and Kacprzyk and Zadrożny [19] which is moderator run with a special "super-agent", a moderator, and its essence is shown in Fig. 1.

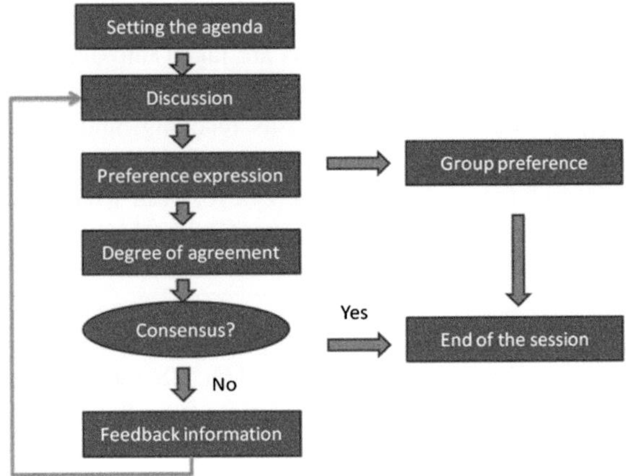

Fig. 1 The essence of a moderator run consensus reaching process with the use of an additional feedback information

We are here mainly concerned with some additional clues, hints, suggestions, etc. that are expressed as some additional indicators to support the moderator—who may use persuasion, suggestions, etc.—to effectively and efficiently change the preferences of some agents with respects to some pairs of options to attain a (sufficient) degree of consensus.

As suggested in our previous works [4, 5] our main concern is to enhance and accelerate the consensus reaching process within the framework mentioned above. Here, we employ in a synergistic way concepts of natural language based knowledge representation and the TOPSIS method [19] based on the aggregation of the "closeness to the ideal" and "farness from the anti-ideal"—cf. [2, 15, 16]. Basically, we define the very concepts of the ideal and anti-ideal with respect to some agents that, roughly speaking, then serve as some reference points, and then proceed to the analysis of what is the best choice of pairs of options the preferences between which should be changed to increase the degree of consensus. The choice of those options is guided by some indicators proposed by Kacprzyk and Zadrożny [22], represented in a human consistent way as linguistic summaries.

3 A Fuzzy Majority Based Concept of a Degree of Consensus

We have a finite set of $m \geq 2$ agents (individuals, decision makers, experts, …), $E = \{e_1, e_2, \ldots, e_m\}$, and a set of $n \geq 2$ options (alternatives, issues, …), $S = \{s_1, s_2, \ldots, s_n\}$. Each agent $e_k \in E$ expresses his/her testimony as to the particular pairs of options from S as an individual *fuzzy preference relation* R_k defined in $S \times S$ [11] which is given by its membership function $\mu_{R_k}: S \times S \to [0, 1]$ such that $\mu_{R_k}(s_i, s_j) \in [0, 1]$ expresses the preference degree of option s_i over option s_j; usually, it is assumed that $\mu_{R_k}(s_i, s_j) > 0.5$ denotes that option s_i is preferred over option

s_j while $\mu_{R_k}(s_i, s_j) < 0.5$ denotes that option s_j is preferred over s_i, with a corresponding degree, and $\mu_{R_k}(s_i, s_j) = 0.5$ means the indifference; this membership function may be conveniently represented as a preference matrix $[r_{ij}^k]$, and it is usually assumed that $r_{ij}^k + r_{ji}^k = 1$, $r_{ii}^k = 0$, for all i, j, k.

The concept of a consensus degree employed is based on a fuzzy majority due to Kacprzyk [12] which is equated with a fuzzy linguistic quantifier (dealt with using Zadeh's calculus of linguistically quantified propositions [29]). The linguistic quantifier, e.g. *most, at least half; almost all,* is assumed to be a fuzzy set in [0, 1], and we use the *relative* fuzzy quantifiers as they are better suited for the *fuzzy majority* representation.

A *linguistically quantified proposition*, such as "most individuals are satisfied", can be written as [29]:

$$Qy's\ are\ F \tag{1}$$

where Q is a linguistic quantifier (e.g., most), $Y = \{y\}$ is a set of objects (e.g., agents) and F is a property (e.g., satisfied).

The fuzzy linguistically quantifier Q is assumed to be a fuzzy set in [0, 1] as, e.g., Q = "*most*", given by

$$\mu_{\text{"most"}}(x) = \begin{cases} 1 & for \quad x > 0.8 \\ 2x - 0.6 & for\ 0.3 \le x \le 0.8 \\ 0 & for \quad x < 0.3. \end{cases} \tag{2}$$

Property F is defined as a fuzzy set in the set of objects Y, and if $Y = \{y_1, \ldots, y_p\}$, then the truth value (degree of truth) of the proposition y_i *is* F is $truth(y_i\ is\ F) = \mu_F(y_i)$, $i = 1, \ldots, p$. The degree of truth of the linguistically quantified proposition (1), *Qy's are F*, is now calculated in two steps:

$$z = \frac{1}{p} \sum_{i=1}^{p} \mu_F(y_i) \tag{3}$$

$$truth(Qy's\ are\ F) = \mu_Q(z) \tag{4}$$

A soft degree of consensus introduced by Kacprzyk and Fedrizzi [15, 16] is meant as the degree to which, for instance: "*most of agents agree with their preferences to most of the options*". Traditionally, consensus is meant to occur only when "all the agents agree with their preferences to all of the options" but such a "full and unanimous agreement" is unrealistic in practice, because agents usually reveal significant differences in their viewpoints, flexibility, aversion to change opinions, etc. [5].

This soft degree of consensus, which is employed here, is derived in three steps [10, 15, 16]:

(1) for each pair of agents we find a degree of agreement as to their preferences between all the pairs of options,

(2) we aggregate these degrees to derive a degree of agreement of each pair of individuals as to their preferences between Q_1 (a linguistic quantifier as, e.g., “most”) pairs of options,

(3) we aggregate these degrees to obtain a degree of agreement of Q_2 (a linguistic quantifier similar to Q_1) pairs of individuals as to their preferences between Q_1 pairs of options, and this is meant to be the *degree of consensus*.

Specifically, the degree of a *strict agreement* between agents e_{k_1} and e_{k_2} as to their preferences between options s_i and s_j, is

$$v_{ij}(k_1, k_2) = \begin{cases} 1 & \textit{if} \quad r_{ij}^{k_1} = r_{ij}^{k_2} \\ 0 & \textit{otherwise} \end{cases} \tag{5}$$

where, $k_1 = 1, \ldots, m-1,\ k_2 = k_1 + 1, \ldots, m,\ i = 1, \ldots, n-1,\ j = i+1, \ldots, n.$

The agreement in preferences concerning some options may be more important for the consensus than in case of some other options. Here, the relevance of options is assumed to be a fuzzy set in the set of options, B, such that $\mu_B(s_i) \in [0,\ 1]$ is a degree of relevance of option $s_i \in S$, from 0 for fully irrelevant to 1 for fully relevant, through all intermediate values. The relevance b_{ij} of a pair of options, $(s_i, s_j) \in S \times S$, may be defined, for instance, as

$$b_{ij}^B = \frac{1}{2}[\mu_B(s_i) + \mu_B(s_j)] \tag{6}$$

for each i,j, where $i \neq j$. Evidently $b_{ij}^B = b_{ji}^B$, for each i,j.

Then, the degree of agreement between agents e_{k_1} and e_{k_2} as to their preferences between *all* pairs of relevant options is:

$$v^B(k_1, k_2) = \frac{\sum_{i=1}^{n-1} \sum_{j=i+1}^{n} v_{ij}(k_1, k_2) * b_{ij}^B}{\sum_{i=1}^{n-1} \sum_{j=i=1}^{n} b_{ij}^B} \tag{7}$$

where $*$ denotes a t-norm, for instance the minimum.

Next, the degree of agreement between agents e_{k_1} and e_{k_2} as to their preferences between Q_1 pairs of relevant options is:

$$v_{Q_1}^B(k_1, k_2) = \mu_{Q_1}(v^B(k_1, k_2)) \tag{8}$$

The degree of agreement of *all* pairs of agents as to their preferences between Q_1 pairs of relevant options is:

$$v_{Q_1}^B = \frac{2}{m(m-1)} \sum_{k_1=1}^{m-1} \sum_{k_2=k_1+1}^{m} v_{Q_1}^B(k_1, k_2) \tag{9}$$

Finally, according to the third step, the degree of agreement of Q_2 pairs of agents as to their preferences between Q_1 pairs of relevant options, called the degree of (strict) consensus, is:

$$con_{E,S}^B(Q_1, Q_2) = \mu_{Q_2}(v_{Q_1}^B) \tag{10}$$

We can relax the strict agreement (5) by a *sufficient agreement* (at least to a degree $\alpha \in [0, 1]$) of agents e_{k_1} and e_{k_2} as to their preferences between options s_i and s_j, defined by:

$$v_{ij}^{\alpha}(k_1, k_2) = \begin{cases} 1 & if \quad |r_{ij}^{k_1} - r_{ij}^{k_2}| \leq 1 - \alpha \\ 0 & otherwise \end{cases} \tag{11}$$

and then, following the same steps as for the strict consensus, we obtain the degree of sufficient agreement of Q_2 pairs of individuals as to their preferences between Q_1 pairs of relevant options, i.e. a degree of sufficient consensus, denoted by

$$con_{E,S}^{B,\alpha}(Q_1, Q_2) = \mu_{Q_2}(v_{Q_1}^{B,\alpha}) \tag{12}$$

For some other possible extensions of (5), which lead to other versions of a degree of consensus, cf. [15, 16]. In what follows, we will sometimes drop the relevance B for the sake of the simplicity of the notation.

4 Some Additional Meta-Information for Helping to Run the Consensus Reaching Session

Obviously, as in the case of decision support in general, it might be important that the moderator be supported by some additional information measuring the current state of the agreement, pointing out some difficulties in reaching consensus, showing which preference matrix/matrices are promising as possible candidates for consensus, etc. Clearly, since the process of consensus reaching is heavily human centered and involves human beings in our context, it would be good to have that additional information, hints, clues, etc. expressed in natural language, notably as linguistic data summaries as proposed in a series of papers by Kacprzyk and Zadrożny [19–22, 24].

Among many possible linguistic summaries proposed for that purpose by Kacprzyk and Zadrożny [19–22, 24], we employ here the following ones which have been suggested by Gołuńska et al. [9] as intuitively appealing and which can contribute to an effective and efficient running of the consensus reaching process:

(1) *The response to omission* of an agent $e_k \in E$, $RTO(k) \in [-1,\ 1]$,which is the difference between the consensus degree for the whole group (10) and the consensus degree for the group without agent e_k:

$$RTO(k) = con_{E,S}(Q_1, Q_2) - con_{E-\{e_k\},S}(Q_1, Q_2) \tag{13}$$

so that it yields a degree of influence of a specific agent, e_k, on the degree of agreement of the group, from -1, for a totally negative influence, through 0 for a lack of influence, to 1 for a totally positive influence; these values are theoretically possible but do not happen in practice.

(2) *The personal consensus degree* of an agent $e_k \in E$, $PCD(k) \in [0, 1]$, is the truth value of the following proposition:

Preferences of agent e_k as to the Q_1 (e.g. most) pairs of options are in agreement with the preferences of Q_2 (e.g. most) agents

which may be written as:

$$PCD(k_1) = \mu_{Q_2}[\frac{1}{(m-1)} \sum_{k_1=1, k_2 \neq k_1}^{m} v^B_{Q_1}(k_1, k_2)] \tag{14}$$

where $v^B_{Q_1}(k_1, k_2)$ is given by (8), which takes values from 0 for an agent who is the most isolated with his opinion, to 1 for an agent whose preference is shared by most of agents; through all intermediate values.

(3) *The option consensus degree* for option $s_i \in S$, $OCD(s_i) \in [0, 1]$, is the degree of truth value of the statement:

Q (e.g., most) pairs of agents agree in their preferences with respect to option s_i

which may be written as follows: we first calculate:

$$s_i(k_1, k_2) = \frac{1}{n-1} \sum_{j=1, j \neq i}^{n-1} v_{ij}(k_1, k_2) \tag{15}$$

for all $k_1 = 1, \ldots, m-1,\ k_2 = k_1 + 1, \ldots, m,$ where $v_{ij}(k_1, k_2)$ is given as (5) and (11) or any other suitable version, cf. [15, 16]; basically, $s_i(k_1, k_2)$ may be viewed as the mean agreement in preferences of the pair of individuals k_1 and k_2 with respect to option s_i in their testimonies (preferences between options); then we obtain

$$OCD(s_i) = \mu_Q[\frac{2}{m(m-1)} \sum_{k_1=1}^{m-1} \sum_{k_2=k_1+1}^{m} s_i(k_1, k_2)] \tag{16}$$

which takes values from 0, for substantially different preferences with respect to option s_i in the testimonies (preferences between pairs of options) of most agents; to 1 for the opposite case, i.e. their agreement as to the preferences with respect to option s_i; notice that in the latter case some options, on which there is the above agreement, can be omitted from a further discussion.

(4) *The response to exclusion* of option $s_i \in S$, $RTE(s_i) \in [-1,\ 1]$, is the difference between the consensus degree for the whole set of options (10) and for the set without option s_i:

$$RTE(s_i) = con_{E,S}(Q_1, Q_2) - con_{E,S-\{s_i\}}(Q_1, Q_2\}) \tag{17}$$

so that it determines the influence of a given option on the consensus degree.

These are the discussion guidance indicators which will be used in this paper, for many other ones, cf. Kacprzyk and Zadrożny [17–22, 24], Kacprzyk et al. [25], Herrera-Viedma et al. [9] or Gołuńska et al. [8], etc. All of them can help support the moderator to properly choose the most promising agents and pairs of options to work on.

5 The Use of the Concept of an Ideal and an Anti-Ideal

In this paper we present an extension of our proposal (cf. Gołuńska et al. [8]) of using the concepts of an ideal and an anti-ideal, and then to use them in the context of TOPSIS [19], to improve the running of the process of consensus reaching. The ideal and anti-ideal may concern both the agents and options.

The idea is as follows. The process of consensus reaching usually starts with the agents' preference relations on the set of options S which are different, and should be made possibly similar (reach a consensus) step by step by changes of preferences by particular agents. Since it is obvious that human agents may be reluctant to change their opinions (though they should to do so to be able to reach consensus), any change involves some "cost" which should be taken into account as proposed by Gołuńska and Kacprzyk [6], and Gołuńska et al. [8, 9]. To be more specific, for simplicity and making the approach operational, the preferences over the set of options are assumed to be quantified using the values of 0, 0.1, 0.2, …, 0.9, 1, and each change of preferences by 0.1 is assumed to imply a unit cost.

A simple and intuitively appealing reasoning, in the context of our additional indicators (13)–(17), can be as follows. We assume as the *ideal agent* the one the testimonies (preferences over the set of options) of whom are in the highest possible agreement with those of other agents. That is, a natural choice can here be that agent, $e_k \in E$, for whom the personal consensus degree, $PCD(k)$, i.e. the truth value (14) of "Preferences of agent e_k as to the $Q_{\neg\neg 1}$ (e.g. most) pairs of options are in agreement with the preferences of $Q_{\neg\neg 2}$ (e.g. most) agents" is the highest. Such an agent would usually best contribute to the consensus, as he or she has the highest agreement with the group with respect to testimonies. On the other hand, since the very definition of the *PCD(.)* does not exclude some differences of the particular agent's testimonies with other members of the group, in the consensus reaching process there may occur some changes of opinions, including those of the ideal agent defined as above, but one can naturally expect that—since that agent's testimonies are the closest to other members of the group—the number of those changes will be the minimal, and therefore the cost of them will be the lowest one among all agents.

And, analogously, an agent with the minimum value of the *PCD* indicator is assumed to be an *anti-ideal agent* as the degree of truth of the statement "his or her preferences as to the $Q_{\neg\neg 1}$ (e.g. most) pairs of options are in agreement with the preferences of $Q_{\neg\neg 2}$ (e.g. most) agents" is the lowest. Therefore, to arrive at a consensus (or, more generally, a better agreement with the entire group) more changes of preferences are needed and the corresponding cost is higher.

Formally, the ideal and anti-ideal agents, denoted by $e_{k^*} \in E$ and $e_{k^-} \in E$, respectively, can be therefore written as:

$$e_{k^*} := \arg \max_{e_k \in E} (PCD(k)) \tag{18}$$

$$e_{k^-} := \arg \min_{e_k \in E} (PCD(k)) \tag{19}$$

Then, the individual fuzzy preference relation matrices of the ideal agent, e_{k^*}, and the anti-ideal agent, e_{k^-}, are denoted as $[r^*_{ij}]$, and $[r^-_{ij}]$, respectively. Notice that the ideal and anti-ideal agents are real agents from within the group though there also exist some approaches in which a non-existing agent is assumed as the ideal and/or anti-ideal agent.

The above reasoning was performed at the level of agents, using the *PCD* (14) indicator. Now, we will show how to employ the other indicators.

Now, since usually, there exist a discrepancy of testimonies between the agents, it might be good to attain such a change of testimonies that they be closer to the ideal agent which somehow epitomizes the consensus as his or her testimonies best reflect what the entire group of agents think.

A plausible approach might here be as follows. Having found the ideal and anti-ideal agent, we should make a further step since the testimonies of the agents may differ from those of the ideal and anti-ideal agent. Namely, we should find options to be presented by the moderator to the group of agents in question which should be most effective and efficient in terms of a possible increase of the degree of consensus. There are two possibilities: first, to present to each agent a different option such that his or her testimonies (preferences) with respect to that option should be individually changed. It is easy to imagine that this may be not efficient enough and it would be better to determine such an option from the point of view of the whole group, in the spirit of the option consensus degree, $OCD(s_i) \in [0, 1]$, for each option $s_i \in S$, which stands for the degree of truth of the statement: "Q (e.g., most) pairs of agents agree in their preferences with respect to option s_i". Clearly, it makes sense to take into account such an option if we go for the taking into account of one option for the entire group, that is we somehow extend the softly defined group of agents in question expressed by "most" to the entire group of agents; this is done for making the approach operational because it would be much more difficult and less intuitively appealing to try to apply the above very plausible rule to some imprecisely defined group of agents exemplified by "most", "almost all", etc. The option for which the truth value of that statement, i.e. "Q (e.g., most) pairs of agents agree in their preferences with respect to option s_i", will be the highest will be our *ideal option* while the one for which the truth value will be the lowest, will be the *anti-ideal* option. It is easy to see that the concept of an ideal option reflects the fact that preference of most agents with respect to that option are the closest, on the average to those of the ideal agent, and analogously for the anti-ideal option.

We implement the above plausible reasoning in the following simple way, which corresponds to the very essence of the well-known TOPSIS [19] method of multicriteria decision making that has proven to work well in our case.

For each agent e_k, $k = 1, \ldots, m$, and each option, $s_i, = 1, 2, \ldots, n$, we calculate the distance of the agent's preference matrix from, first, that of the ideal agent, and second, from that of the anti-ideal agent. That is, using for simplicity the Euclidean distance, we compute for the ideal agent and for all agents $e_k \in E$ and all options $s_i \in S$:

$$d^*_{s_i}(e_k) = \frac{1}{n-1}\sqrt{\sum_{j=1, j\neq i}^{n} (r^k_{ij} - r^*_{ij})^2}$$

and then for each option $s_i \in S$ we obtain:

$$d^*_{s_i} = \frac{1}{m}\sum_{k=1}^{m} d^*_{s_i}(e_k) = \frac{1}{m}\sum_{k=1}^{m} (\frac{1}{n}\sqrt{\sum_{j=1}^{n} (r^k_{ij} - r^*_{ij})^2}) \tag{20}$$

which yields the goodness of option s_i as a candidate for proposing it to the whole group for possibly changing its preferences of other options with respect to s_i so that the particular agents could get closer in their preferences to the ideal agent, and hence possibly increasing the consensus degree. The higher the value of $d^*_{s_i}$ the better candidate for the discussion s_i is.

And similarly, for the to the anti-ideal agents we compute for all agents $e_k \in E$ and all options $s_i \in S$:

$$d^-_{s_i}(e_k) = \frac{1}{n}\sqrt{\sum_{j=1}^{n} (r^k_{ij} - r^-_{ij})^2}$$

and

$$d_{s_i}^{-} = \frac{1}{m}\sum_{k=1}^{m} d_{s_i}^{-}(e_k) = \frac{1}{m}\sum_{k=1}^{m}\left(\frac{1}{n}\sqrt{\sum_{j=1}^{n}(r_{ij}^{k} - r_{ij}^{-})^2}\right) \tag{21}$$

which yields the goodness of option s_i as a candidate for proposing it to the whole group for possibly changing its preferences of other options with respect to s_i so that the particular agents could get closer in their preferences to the anti-ideal agent, and hence possibly increasing the consensus degree. This time, the lower the value of $d_{s_i}^{-}$ the better candidate for the discussion s_i is.

Finally, following the idea of TOPSIS, we compute, for each option s_i, with respect to the ideal agent, e_{k^*}, and the anti-ideal agent, e_{k^-}:

$$C_{s_i}^{*} = \frac{d_{s_i}^{-}}{d_{s_i}^{*} + d_{s_i}^{-}}\ ;\ i = 1, \ldots, n \tag{22}$$

which is an aggregation of the "closeness to the ideal" and "farness from the anti-ideal".

Therefore, if option s_i is the ideal option, in the sense that for it (22) takes on the highest value, denoted by s_i^*, then $C_{s_i}^* = 1$, while if s_i^- is the anti-ideal solution, in the sense that for it (22) takes on the lowest value, then $C_{s_i}^* = 0$. Therefore, the closer the value of $C_{s_i}^*$ to 1, the closer option s_i to the ideal solution s_i^*.

In this paper, in the definitions given, we have practically used the ideas behind the two additional pieces of information only, i.e. PCD (14) and OCD (16). but the additional measures mentioned, i.e. the response to omission RTO (13) and the response to exclusion RTE (17) can be employed as additional terms in (18) and (19). One can also use other additional measures proposed by Kacprzyk and Zadrożny [22] or Kacprzyk et al. [25].

Finally, let us mention that the method of using additional information proposed may essentially be meant as to provide an enhanced feedback mechanism that should help the moderator as shown in Fig. 2.

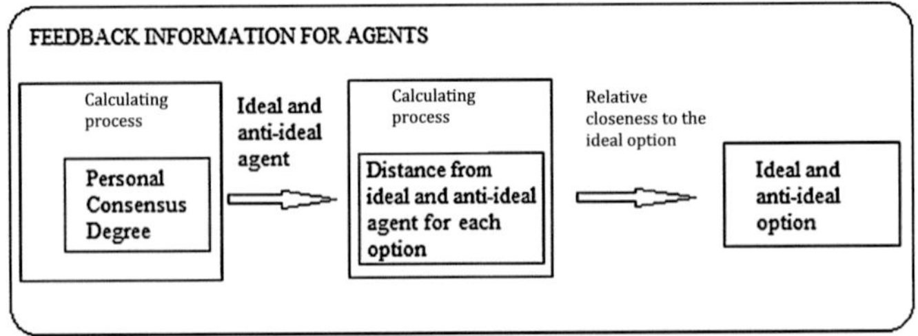

Fig. 2 An extension of the feedback mechanism

6 A Numerical Example

We will now solve the simple example given in Gołuńska et al. [9]. We have a group of 5 agents and 4 options considered, and their initial individual fuzzy preference relations matrices (the irrelevant left lower triangles are omitted as the reciprocity of the preference relations is assumed) are:

$$[r_{ij}^1] = \begin{bmatrix} - & 0.9 & 0.9 & 1.0 \\ & - & 0.8 & 0.7 \\ & & - & 0.7 \\ & & & - \end{bmatrix} [r_{ij}^2] = \begin{bmatrix} - & 0.7 & 1.0 & 1.0 \\ & - & 0.8 & 0.9 \\ & & - & 0.5 \\ & & & - \end{bmatrix} [r_{ij}^3] = \begin{bmatrix} - & 0.9 & 0.8 & 0.6 \\ & - & 0.6 & 0.3 \\ & & - & 0.3 \\ & & & - \end{bmatrix}$$

$$[r_{ij}^4] = \begin{bmatrix} - & 0.9 & 0.5 & 1.0 \\ & - & 0.0 & 0.4 \\ & & - & 1.0 \\ & & & - \end{bmatrix} [r_{ij}^5] = \begin{bmatrix} - & 0.9 & 0.8 & 0.9 \\ & - & 0.2 & 0.5 \\ & & - & 0.9 \\ & & & - \end{bmatrix}$$

The degree of the sufficient consensus (11), with $Q_1 = Q_2 =$ "most' given by (12), and with $\alpha = 0.8$ is (the relevance of all options, B, is assumed to be equal 1 and thus B is not mentioned in the following notation):

$$con^{\alpha}(Q_1, Q_2) = 0.43$$

This value is considered too low to be acceptable and we resort to the method proposed in this paper.

The ideal agent corresponds to the agent with the highest PCD value. The values of $PCD(k)$, k = 1, 2, …, 5, for the particular agents, are shown rank ordered in Table 1. It can be seen that *agent* e_5 is the ideal agent, because it has the highest PCD value, while *agent* e_4 is the anti-ideal one because it has the lowest one.

From (22) we obtain the values of of $C_{s_i}^*$, i = 1, 2, …, 4, as given in Table 2. Option s_1 is therefore the best one, and option s_4 is the worst one.

The moderator suggests that the preferences of *all agents* as to the preferences with respect to the *ideal option* s_1 should be changed to be equal to the preferences expressed by the *ideal agent* e_5. There are 8 changes needed, i.e. it costs 8 cost units.

Table 1 Values of *PCD* for each agent

PCD(k)	Value
PCD(5)	0.8
PCD(1)	0.6
PCD(2)	0.5
PCD(3)	0.2
PCD(4)	0

Table 2 Average closeness to the preferences of the ideal agent

$C^*_{s_i}$	Value
C^*_{s1}	0.62
C^*_{s3}	0.58
C^*_{s2}	0.57
C^*_{s4}	0.55

Table 3 Values of *PCD* for each agent

PCD(k)	Value
PCD(3)	1
PCD(1)	0.97
PCD(5)	0.97
PCD(2)	0.8
PCD(4)	0.67

The newly obtained degree of consensus is now equal to:

$$con^{\alpha}(Q_1, Q_2) = 0.88.$$

and suppose that it is still considered not high enough and the second round is run as before.

Table 3 shows that now *agent* e_3 is the ideal agent, while *agent* e_4 is still the anti-ideal one.

Then, for a lack of space, the respective table, similar to Table 2, will not be shown, but the average closeness the respective preferences of the ideal agent indicates *option* s_2 as the one which should be used now. Because of a high degree of consensus the moderator starts to persuade now the *anti-ideal agent* to change his preferences concerning the *ideal option* s_2. The new values of his or her preferences should be the same as those expressed by the *ideal agent* e_3. It turns out that the change in the anti-ideal agents' preferences:

$$r^{-}_{23} = 0.0 \rightarrow 0.6$$

cause the acceptable value of the group consensus degree, which is now:

$$con^{\alpha}(Q_1, Q_2) = 0.93.$$

with the cost equal to 6.

As it can be seen the use of the concept of an ideal and anti-ideal agent and option can considerably help attain a higher value of consensus by better supporting the moderator. Moreover, a possible inclusion of the response to omission RTO (13) and the response to exclusion RTE (17) can be employed for a further improvement but this will be considered in next works.

7 Concluding Remarks

In the paper we have extended our moderator run consensus reaching model (cf. proposed by Gołuńska et al. [8]) by including, first, some new information that may be useful for the moderator to run the consensus reaching session which is given as linguistic summaries, in the sense of Yager, i.e. represented by linguistically quantified propositions with their associated degrees of truth, and second, a novel synergistic combination of the ideal and anti-ideal point (agents and options) which are then used in a TOPSIS like procedure as proposed in a simpler setting by Gołuńska et al. [8]. The results are promising in terms of both the value of a consensus degree obtained and the speed and cost of attaining that value, i.e. the number of necessary changes of preferences of individuals.

As a promising direction for a further research, the use of more sophisticated discussion guiding linguistic summaries can be considered as proposed by Kacprzyk and Zadrożny [22], Kacprzyk et al. [25], etc.

Acknowledgments This work is partially supported by the Foundation for Polish Science under the "International PhD Projects in Intelligent Computing" financed from the European Union within the Innovative Economy Operational Programme 2007–2013 and European Regional Development Fund, and partially by the National Science Centre under Grant No. UMO–2012/05/B/ST6/03068.

References

1. Anagnostopoulos, K., Doukas, H., Psarras, J.: A linguistic multicriteria analysis system combining fuzzy sets theory, ideal and anti-ideal points for location site selection. Expert Syst. Appl. **35**(4), 2041–2048 (2008)
2. Carlsson, C., Fedrizzi, M., Fuller, R.: Group decision support systems. In: Carlsson, C., Fedrizzi, M., Fuller, R. (eds.) Fuzzy Logic in Management, **66**, pp. 57–125. Springer Science, Berlin (2004)
3. Fedrizzi, M., Fedrizzi, M., Marques Pereira, R.A.: Consensus modelling in group decision making: a dynamical approach based on Zadeh's fuzzy preferences. In: Seising, R., Trillas, E., Moraga, C, Termini, S. (eds.) On Fuzziness (Homage to Lotfi A. Zadeh), pp. 165–170. Springer, Heidelberg (2013)
4. Fedrizzi, M., Kacprzyk, J., Zadrożny, S.: An interactive multi-user decision support system for consensus reaching process using fuzzy logic with linguistic quantifiers. Decis. Support Syst. **4**(3), 313–327 (1988)
5. Gołuńska, D, Hołda, M.: The need of fairness in the group consensus reaching process in a fuzzy environment. Tech. Trans. Autom. Control, vol. 1-AC/2013, pp. 29–38 (2013)
6. Gołuńska, D., Kacprzyk, J.: The conceptual framework of fairness in consensus reaching process under fuzziness. In: Proceedings of the 2013 Joint IFSA World Congress NAFIPS Annual Meeting, pp. 1285–1290. Edmonton, Canada, 24–28 June 2013
7. Gołuńska, D., Kacprzyk, J., Herrera-Viedma, E.: Modeling consensual process for group decision making problem with reference to various attitudes of advising decision makers. In: Proceedings of IEEE Intelligent Systems 2014. Springer, Heidelberg (2014) (in press)
8. Gołuńska, D., Kacprzyk, J., Zadrożny, S.: A model of efficiency-oriented group decision and consensus reaching support system in a fuzzy environment: In: Proceedings of the 15th

International Conference on Information Processing and Management of Uncertainty in Knowledge-Based Systems, pp. 424–433. IPMU-2014, (2014)
9. Gołuńska, D., Kacprzyk, J., Zadrożny, S.: On efficiency-oriented support of consensus reaching in a group of agents in a fuzzy environment with a cost based preference updating approach. In: Proceeding of SSCI-2014, Orlando, FL, USA, IEEE Press (2014) (forthcoming)
10. Herrera-Viedma, E., Cabrerizo, F.J., Kacprzyk, J., Pedrycz, W.: A review of soft consensus models in a fuzzy environment. Inf. Fusion **17**, 4–13 (2014)
11. Herrera-Viedma, E., García-Lapresta, J.-L., Kacprzyk, J., Fedrizzi, M., Nurmi, H., Zadrozny, S. (eds.): Consensual Processes. Studies in Fuzziness and Soft Computing 267. Springer, Berlin (2011)
12. Kacprzyk, J.: Group decision making with a fuzzy linguistic majority. Fuzzy Sets Syst. **18**(2), 105–118 (1986)
13. Kacprzyk, J.: Multistage Decision Making in Fuzzy Conditions (in Polish). PWN, Warszawa (1983)
14. Kacprzyk, J.: Neuroeconomics: yet another field where rough sets can be useful?. In: Chan, C.-C., Grzymała-Busse, J.W., Ziarko, W.P. (eds.) Rough Sets and Current Trends in Computing. LNAI 5306, pp. 1–12. Springer, Berlin (2008)
15. Kacprzyk, J., Fedrizzi, M.: A 'soft' measure of consensus in the setting of partial (fuzzy) preferences. Eur. J. Oper. Res. **34**, 315–325 (1988)
16. Kacprzyk, J., Fedrizzi, M.: A 'human-consistent' degree of consensus based on fuzzy logic with linguistic quantifiers. Math. Soc. Sci. **18**, 275–290 (1989)
17. Kacprzyk, J., Fedrizzi, M., Nurmi, H.: Soft degrees of consensus under fuzzy preferences and fuzzy majorities. In: Kacprzyk, J., Nurmi, H., Fedrizzi, M. (eds.) Consensus under Fuzziness, pp. 55–83. Kluwer Academic Publishers, Boston (1996)
18. Kacprzyk, J., Zadrożny, S.: On the use of fuzzy majority for supporting consensus reaching under fuzziness. In: Proceedings of FUZZ-IEEE'97—Sixth IEEE International Conference on Fuzzy Systems, vol. 3, pp. 1683–1988 (1997)
19. Kacprzyk, J., Zadrożny, S.: On a concept of a consensus reaching process support system based on the use of soft computing and Web techniques. In: Ruan, D., Montero, J., Lu, J., Martínez, L., D'hondt, P., Kerre, E.E., (eds.) Computational Intelligence in Decision and Control, pp. 859–864. World Scientific, Singapore (2008)
20. Kacprzyk, J., Zadrożny, S.: Towards a general and unified characterization of individual and collective choice functions under fuzzy and nonfuzzy preferences and majority via the ordered weighted average operators. Int. J. Intell. Syst. **24**(1), 4–26 (2009)
21. Kacprzyk, J., Zadrożny, S.: Soft computing and Web intelligence for supporting consensus reaching. Soft. Comput. **14**(8), 833–846 (2010)
22. Kacprzyk, J., Zadrożny, S.: Supporting consensus reaching processes under fuzzy preferences and a fuzzy majority via linguistic summaries. In: Greco, S., Marques Pereira, R.A., Squillante, M., Yager, R.R., (eds.) Preferences and Decisions, vol. 257, pp. 261–279 (2010)
23. Kacprzyk, J., Zadrożny, S.: Computing with words is an implementable paradigm: fuzzy queries, linguistic data summaries and natural language generation. IEEE Trans. Fuzzy Syst. **18**(3), 461–472 (2010)
24. Kacprzyk, J., Zadrożny, S., Fedrizzi, M., Nurmi, H.: On group decision making, consensus reaching, voting and voting paradoxes under fuzzy preferences and a fuzzy majority: a survey and some perspectives. In: Bustince, H., Herrera, F., Montero, J. (eds.) Fuzzy Sets and Their Extensions: Representations, Aggregation and Models, pp. 263–295. Springer, Heidelberg (2008)
25. Kacprzyk, J., Zadrożny, S., Raś, Z.W.: How to support consensus reaching using action rules: a novel approach. Int. J. Uncertainty Fuzziness Knowl.-Based Syst. **18**(4), 451–470 (2010)
26. Kacprzyk, J., Zadrożny, S., Wilbik, A.: Linguistic summarization of some static and dynamic features of consensus reaching. In: Reusch, B., (ed.) Computational Intelligence, Theory and Applications, pp.19–28. Springer, Berlin (2006)

27. Szmidt, E., Kacprzyk, J.: A consensus-reaching process under intuitionistic fuzzy preference relations. Int. J. Intell. Syst. **18**(7), 837–852 (2003)
28. Turban, E., Aronson, J.E., Liang, T.P.: Decision Support Systems and Intelligent Systems, 6th edn. Prentice Hall, Upper Saddle River (2005)
29. Zadeh, L.: A computational approach to fuzzy quantifiers in natural languages. Comput. Math Appl. **9**, 149–184 (1983)

Interval Type-2 Fuzzy System Design Based on the Interval Type-2 Fuzzy C-Means Algorithm

Elid Rubio, Oscar Castillo and Patricia Melin

Abstract In this work, the Interval Type-2 Fuzzy C-Means (IT2FCM) algorithm is used for the design of Interval Type-2 Fuzzy Inference Systems using the centroids and fuzzy membership matrices for the lower and upper bound of the intervals obtained by the IT2FCM algorithm in each data clustering realized by this algorithm, and with these elements obtained by IT2FCM algorithm we design the Mamdani, and Sugeno Fuzzy Inference systems for classification of data sets and time series prediction.

1 Introduction

Due to need of finding interesting patterns or groups of data with similar characteristics in a given data set, Clustering algorithms [1, 3, 5, 11, 12, 15] have been proposed to satisfy this need. Currently there are various fuzzy clustering algorithms. The Fuzzy C-Means (FCM) algorithm [1, 12] has been the foundation to developing other clustering algorithms, which have been widely used successfully in pattern recognition [1], data mining [6], classification [8], image segmentation [16, 25], data analysis and modeling [3]. The popularity of the fuzzy clustering algorithms is due to the fact that allow a datum belong to different data clusters into a given data set.

However this method is not able to handle the uncertainty found in a given dataset during the process of data clustering; because of this the FCM was extended to IT2FCM using Type-2 Fuzzy Logic Techniques [10, 14]. This extension of the

E. Rubio · O. Castillo (✉) · P. Melin
Tijuana Institute of Technology, Tijuana, Mexico
e-mail: ocastillo@tectijuana.mx

E. Rubio
e-mail: elid.rubio@hotmail.com

P. Melin
e-mail: pmelin@tectijuana.mx

M. Collan et al. (eds.), *Fuzzy Technology*, Studies in Fuzziness and Soft Computing 335, DOI 10.1007/978-3-319-26986-3_8

FCM algorithm has been applied to the formation of membership functions [4, 7, 20, 21], and classification [2]. In this work the creation of fuzzy systems are presented using the IT2FCM algorithm using the centroids matrices and fuzzy partition for the lower and upper limits of the range, with these matrices obtained using IT2FCM the membership functions for each input and output variable of the fuzzy system and its rules of inference are created.

2 Overview of Interval Type-2 Fuzzy Sets

Type-2 Fuzzy Sets are an extension of the Type-1 Fuzzy Sets proposed by Zadeh in 1975, this extension was designed with the aim of mathematically representing the vagueness and uncertainty of linguistic problems and this way overcome limitations of Type-1 Fuzzy Sets and thereby provide formal tools to work with intrinsic imprecision in different type of problems. Type-2 Fuzzy Sets are able to describe uncertainty and vagueness in information, usually are used to solve problems where the available information is uncertain. These Fuzzy Sets include a secondary membership function to model the uncertainty of Type-1 Fuzzy Sets [10, 14].

A Type-2 Fuzzy set in the universal set X, is denoted as $\tilde{A}$, and can be characterized by a Type-2 fuzzy membership function $\mu_{\tilde{A}} = (x, u)$ as:

$$\tilde{A} = \int_{x \in X} \mu_{\tilde{A}}(x)/x = \int_{x \in X} \left[\int_{u \in J_x} f_x(u)/u \right] /x, J_x \subseteq [0, 1] \tag{1}$$

where J_x is the primary membership function of x which is the domain of the secondary membership function $f_x(u)$.

The shaded region shown in Fig. 1a is usually called footprint of uncertainty (FOU). The FOU of $\tilde{A}$ is the union of all primary membership that are within the lower and upper limit of the interval of membership functions and can be expressed as:

$$FOU(\tilde{A}) = \bigcup_{\forall x \in X} J_x = \{(x, u) | u \in J_x \subseteq [0, 1]\} \tag{2}$$

The lower membership function (LMF) and upper membership function (UMF) are denoted by $\underline{\mu}_{\tilde{A}}(x)$ and $\overline{\mu}_{\tilde{A}}(x)$ are associated with the lower and upper bound of $FOU(\tilde{A})$ respectively, i.e. The UMF and LMF of $\tilde{A}$ are two Type-1 membership functions that bound the FOU as shown in Fig. 1a. By definition they can be represented as:

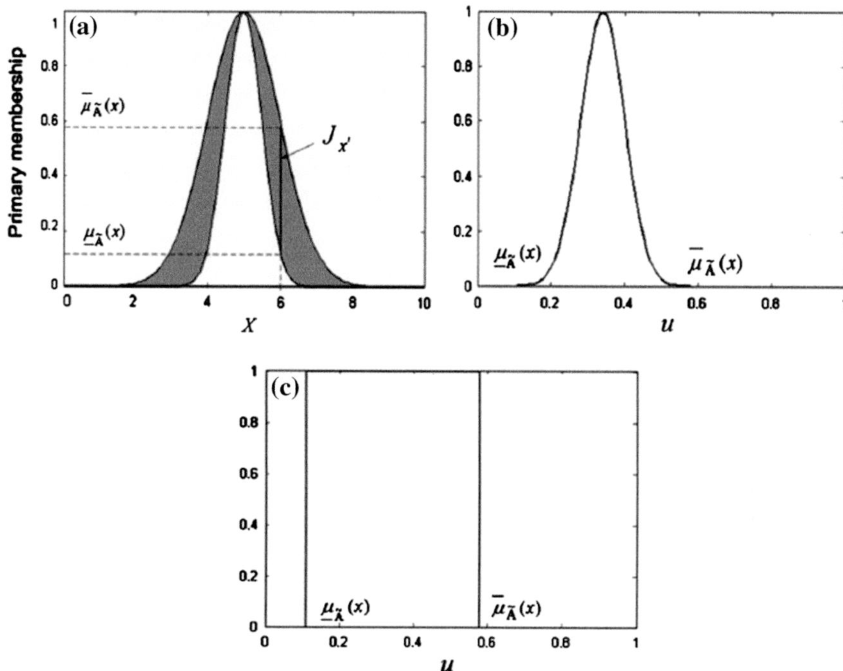

Fig. 1 **a** Type-2 membership function, **b** Secondary membership function and, **c** Interval secondary membership function

$$\underline{\mu}_{\tilde{A}}(x) = \underline{FOU}(\tilde{A}) \forall x \in X \tag{3}$$

$$\bar{\mu}_{\tilde{A}}(x) = \overline{FOU}(\tilde{A}) \forall x \in X \tag{4}$$

The secondary membership function is a vertical slice of $\mu_{\tilde{A}} = (x, u)$ as shown in Fig. 1b. The Type-2 Fuzzy Sets are capable modeling uncertainty, where Type-1 fuzzy sets cannot. The computation operations required by Type-2 fuzzy systems are considerably and, undesirably large, this is due these operations involve numerous embedded Type-2 fuzzy sets which consider all possible combinations of the secondary membership values [10, 14]. However with the aim of reduce the computational complexity was proposed Interval Type-2 Fuzzy Sets, where the secondary membership functions are interval sets expressed as:

$$\tilde{A} = \int_{x \in X} \left[\int_{u \in J_x} 1/u \right] / x \tag{5}$$

Figure 1c shows the membership function of an Interval Type-2 Fuzzy Sets. The secondary memberships are all uniformly weighted for each primary membership of x. Therefore J_x can be expressed as:

$$J_x = \left\{ (x, u) | u \in \left[\underline{\mu}_{\tilde{A}}(x), \overline{\mu}_{\tilde{A}}(x) \right] \right\} \quad (6)$$

Moreover, $FOU(\tilde{A})$ in (2) can be expressed as:

$$FOU(\tilde{A}) = \bigcup_{\forall x \in X} \left\{ (x, u) | u \in \left[\underline{\mu}_{\tilde{A}}(x), \overline{\mu}_{\tilde{A}}(x) \right] \right\} \quad (7)$$

As a result of whole this, the computational complexity using Interval Type-2 Fuzzy Sets is reduced only to calculate simple interval arithmetic.

3 Interval Type-2 Fuzzy C-Means Algorithm

The Interval Type-2 Fuzzy C-Means (IT2FCM) algorithm [4, 7, 20] is an extension of the Fuzzy C-Means (FCM) algorithm, this extension use Type-2 Fuzzy Techniques in combination with the C-Means algorithm, and improvement the traditional FCM, which uses Type-1 Fuzzy Logic [26, 27]. This method is able to handle uncertainty found in a given dataset during the process of data clustering and thereby makes data clustering less susceptible to noise to achieve the goal that data can be clustered more appropriately and more accurately.

The weighting (fuzzification) exponent m in the IT2FCM algorithm is represented by an interval rather than a precise numerical value, i.e. $m = [m_1, m_2]$, where m_1 and m_2 represent the lower and upper limit of the weighting (fuzzification) exponent respectively.

Because the m value is represented by an interval, the fuzzy partition matrix μ_{ij} must be calculated to the interval $[m_1, m_2]$, per this reason μ_{ij} would be given by a membership interval$[\underline{\mu}_i(x_j), \overline{\mu}_i(x_j)]$ where $\underline{\mu}_i(x_j)$ and $\overline{\mu}_i(x_j)$ represents the lower and upper limit of the belonging interval of datum x_j to a clustering v_i, updating the lower an upper limit of the range of the fuzzy membership matrices can be expressed as:

$$\underline{\mu}_i(x_j) = \min \left\{ \left[\sum_{k=1}^{c} \left(\frac{d_{ij}^2}{d_{ij}^2} \right)^{\frac{2}{m_1 - 1}} \right]^{-1}, \left[\sum_{k=1}^{c} \left(\frac{d_{ij}^2}{d_{ij}^2} \right)^{\frac{2}{m_2 - 1}} \right]^{-1} \right\} \quad (8)$$

$$\overline{\mu}_i(x_j) = \max \left\{ \left[\sum_{k=1}^{c} \left(\frac{d_{ij}^2}{d_{ij}^2} \right)^{\frac{2}{m_1 - 1}} \right]^{-1}, \left[\sum_{k=1}^{c} \left(\frac{d_{ij}^2}{d_{ij}^2} \right)^{\frac{2}{m_2 - 1}} \right]^{-1} \right\} \quad (9)$$

The procedure for updating the cluster prototypes in the IT2 FCM algorithm should take into account the degree of belonging interval to calculate the centroids of the fuzzy membership matrix for the lower and upper limit these centroids will be given by the following equations:

$$\underline{v}_i = \frac{\sum_{j=1}^{n} \left(\underline{\mu}_i(x_j)\right)^{m_1} x_j}{\sum_{j=1}^{n} \left(\underline{\mu}_i(x_j)\right)^{m_1}} \tag{10}$$

$$\overline{v}_i = \frac{\sum_{j=1}^{n} \left(\overline{\mu}_i(x_j)\right)^{m_1} x_j}{\sum_{j=1}^{n} \left(\overline{\mu}_i(x_j)\right)^{m_1}} \tag{11}$$

The resulting interval of the coordinates of the centroids positions of the clusters. Type-reduction and defuzzification use Type-2 fuzzy operations. The centroids matrix and the fuzzy partition matrix are obtained by the type-reduction as shown in the following equations:

$$v_j = \frac{\underline{v}_j + \overline{v}_j}{2} \tag{12}$$

$$\mu_i(x_j) = \frac{\underline{\mu}_i(x_j) + \overline{\mu}_i(x_j)}{2} \tag{13}$$

Based on all this description, the IT2 FCM algorithm consists of the following steps:

1. Establish c, m_1, m_2.
2. Initialize fuzzy partition matrices $\underline{\mu}_i(x_j)$ and $\overline{\mu}_i(x_j)$, such that with restriction in:

$$\sum_{i=1}^{c} \underline{\mu}_i(x_j) = 1 \tag{14}$$

$$\sum_{i=1}^{c} \overline{\mu}_i(x_j) = 1 \tag{15}$$

3. Calculate the centroids for the lower and upper fuzzy partition matrix using the Eqs. (10) and (11) respectively.
4. Calculating the update of the fuzzy partition matrices for lower and upper bound of the interval using the Eqs. (8) and (9) respectively.
5. Type reduction of the fuzzy partition matrix and centroid, if the problem requires using the Eqs. (12) and (13) respectively.
6. Repeat steps 3 to 5 until $|J_{\tilde{m}}(t) - J_{\tilde{m}}(t-1)| < \varepsilon$.

This extension on the FCM algorithm is intended to realize that this algorithm is capable of handling uncertainty and is less susceptible noise.

4 Designing Type-2 Fuzzy Inference Systems Using Interval Type-2 Fuzzy C-Means Algorithm

The design of Type-2 Inference Systems is performed taking into account the lower and upper centroid matrices and the lower and upper fuzzy membership matrices generated by the Interval Type-2 Fuzzy C-Means.

But for designing a Fuzzy Inference System is necessary to define input and output variables and rules of inference, the number of input and output variables will be given for the number of dimensions or characteristics of the data input and data output respectively. The number of membership functions for each input and output variable, is given by the number of clusters specified to IT2FCM algorithm for the input and output data clustering.

In creating the membership functions of the input and output variables, the centroid matrices and the fuzzy membership matrices of the lower and upper bounds of the interval are used, in this particular case Type-2 Gaussian membership functions are created this is because that the parameter for this membership functions are center and standard deviation.

The centers for the Gaussian membership functions for the lower and upper bound of the interval are provided by the IT2FCM algorithm. The standard deviation for the lower and upper Gaussian membership functions can be found using the following equations:

$$\underline{\sigma}_i = \frac{1}{n}\sum_{j=1}^{n}\sqrt{-\frac{\left(x_j - \underline{v}_i\right)^2}{2\ln \underline{\mu}_i(x_j)}} \tag{16}$$

$$\overline{\sigma}_i = \frac{1}{n}\sum_{j=1}^{n}\sqrt{-\frac{\left(x_j - \overline{v}_i\right)^2}{2\ln \overline{\mu}_i(x_j)}} \tag{17}$$

As shown in Fig. 2, the calculation of the standard deviation is performed using the matrices of centroids, the fuzzy membership matrices and the dataset to which applied the IT2FCM algorithm to find the above matrices. Once the standard deviation is calculated for each centroid found in the matrices of the lower and upper centroids, one proceeds to create the functions for the lower and upper limit of the interval as shown in Fig. 2, with the following equations:

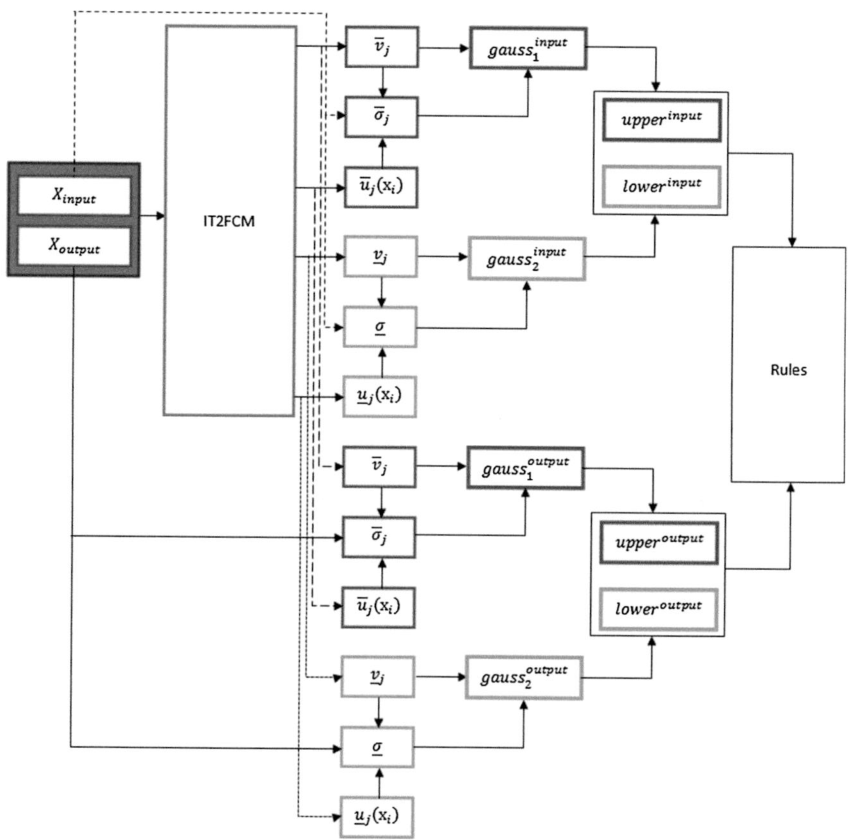

Fig. 2 Block diagram for designing Type-2 fuzzy inference systems using IT2FCM algorithm

$$\underline{gauss} = \min(gauss_1, gauss_2) \tag{18}$$

$$\overline{gauss} = \begin{cases} 1 & \text{if} \min(\underline{v}_i, \overline{v}_i) < x < \max(\underline{v}_i, \overline{v}_i) \\ \max(gauss_1, gauss_2) & \text{otherwise} \end{cases} \tag{19}$$

where

$$gauss_1(x, \underline{\sigma}_i, \underline{v}_i) = e^{-\frac{1}{2}\left(\frac{x - \underline{v}_i}{\underline{\sigma}_i}\right)^2} \tag{20}$$

$$gauss_2(x, \overline{\sigma}_i, \overline{v}_i) = e^{-\frac{1}{2}\left(\frac{x - \overline{v}_i}{\overline{\sigma}_i}\right)^2} \tag{21}$$

Once the input and output variables are established with their respective membership functions are performed the creation of rules as shown in Fig. 2. In this

Table 1 Example of rule creation with IT2FCM

if $input_1$ *is* mf_1 *then* $output_1$ *is* mf_1
if $input_1$ *is* mf_2 *then* $output_1$ *is* mf_2
$\vdots$
if $input_1$ *is* mf_n *then* $output_1$ *is* mf_n

particular case the amount of rules created will depend on the number of functions of membership, that is, if we have a fuzzy system with an input variable and an output variable with three functions of membership in each variable rules would formed as follows, see Table 1.

4.1 Type-2 Fuzzy Systems Designed for Time Series Prediction

In this section we present the structure of the Mamdani and Sugeno T2 FIS designed by the IT2FCM algorithm for the Mackey-Glass time series prediction and the results obtained by T2 FIS designed. In Figs. 3 and 5, the structure of Mamdani and Sugeno are shown respectively of the T2FIS designed for the prediction of time series Mackey- Glass using IT2 FCM algorithm.

From the Mackey-Glass time series 800 pairs of data points were extracted [9, 13, 17–19, 22, 23, 24]. The work of the T2FIS consist in predict $x(t)$ from to a

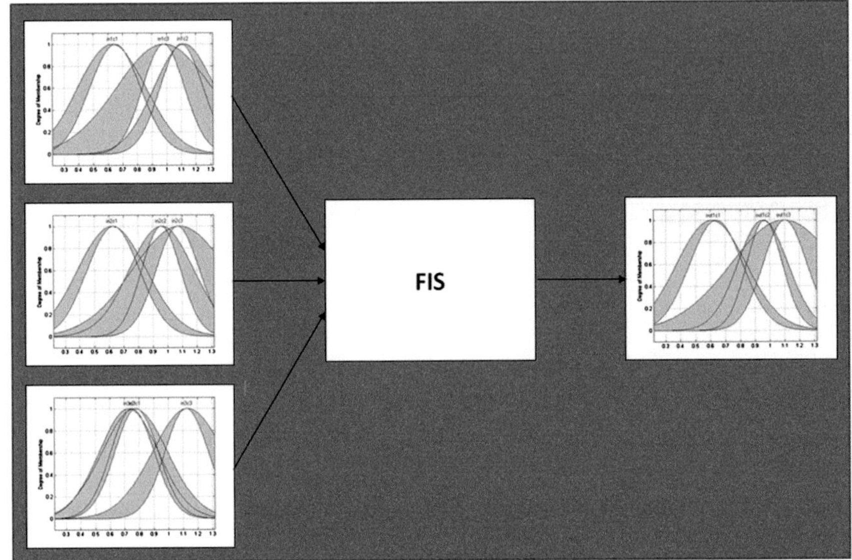

Fig. 3 T2 FIS Mamdani designed by IT2FCM for mackey-glass time series forecast

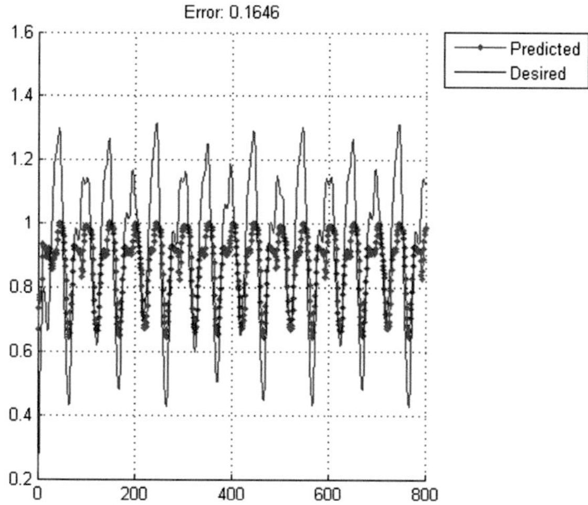

Fig. 4 Results of the Mackey-Glass time series prediction using the Mamdani Type-2 Fuzzy inference system designed by IT2FCM

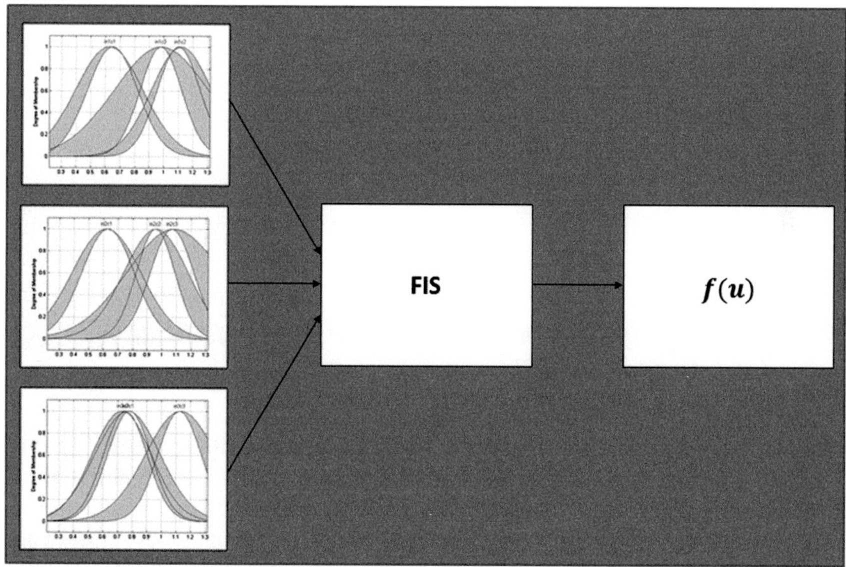

Fig. 5 T2 FIS Sugeno designed by IT2FCM for time series forecast Mackey-Glass

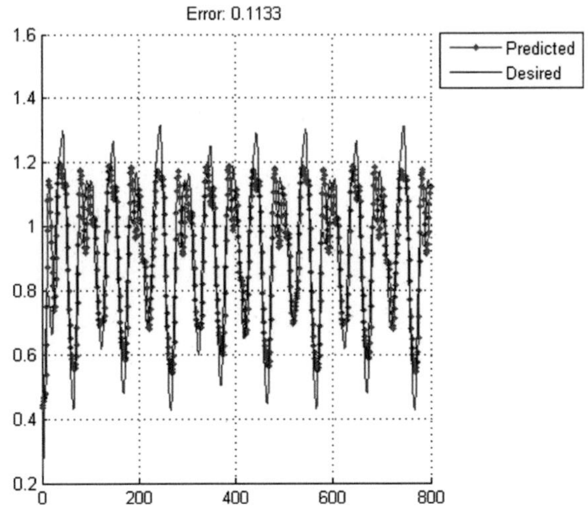

Fig. 6 Result of the Mackey-Glass time series prediction using the Sugeno Type-2 Fuzzy inference system designed by IT2FCM

data set created with three column vectors with 6 periods of delay in each vector of the time series, that is, $x(t-18)$, $x(t-12)$, and $x(t-6)$. Therefore the format of the training data is:

$$[x(t-18), x(t-12), x(t-6)\,; x(t)] \tag{22}$$

where $t = 19$ to 818 and $x(t)$ is the desired prediction of the time series. The first 400 pairs of data are used to create the T2 FIS, while the other 400 pairs of data are used to test the T2 FIS designing. In Figs. 4 and 6, one can observe the result of the prediction of the time series Mackey-Glass using Type-2 Fuzzy Systems designed by IT2FCM algorithm.

4.2 *Type-2 Fuzzy Systems Designed for the Classification of the Dataset Iris Flower*

In this section we present the structure of Mamdani and Sugeno T2 FIS designed by IT2FCM algorithm for classification of the dataset iris flower and the result obtained by T2 FIS designed. In Figs. 7 and 9, are shown the Mamdani and Sugeno structure respectively of the T2FIS designed for classification of the dataset iris flower using IT2 FCM algorithm.

The iris flower dataset consists of 150 samples in each of the 4 characteristics and 3 classes of flowers, i.e., 50 samples per class. To create the T2 FIS, 50 % of the samples are used and the other 50 % of the samples are used to test the fuzzy system created using the IT2FCM algorithm.

In Figs. 8 and 10, one can observe the result of the classification of the iris flower dataset using Type-2 Fuzzy Systems designed by IT2FCM algorithm.

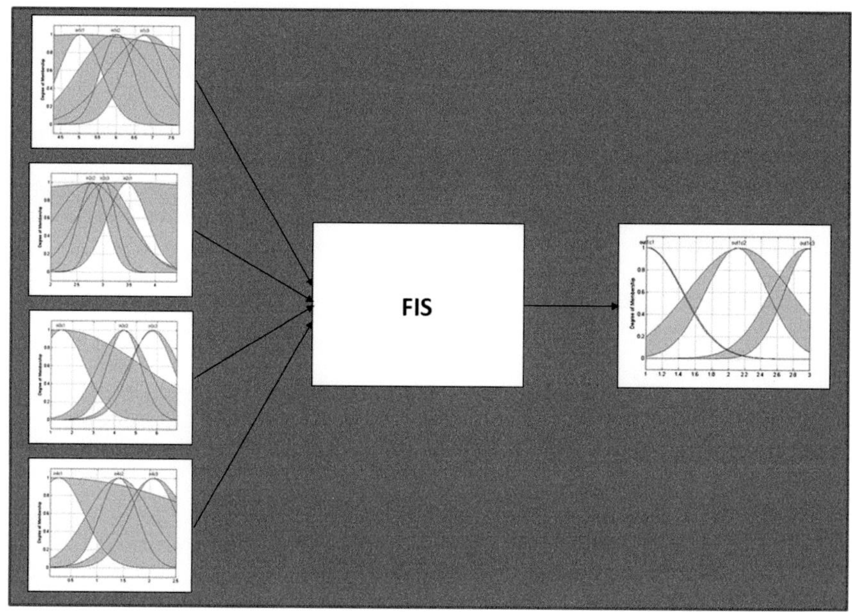

Fig. 7 Mamdani T2 FIS designed by IT2FCM for classification of Iris data set

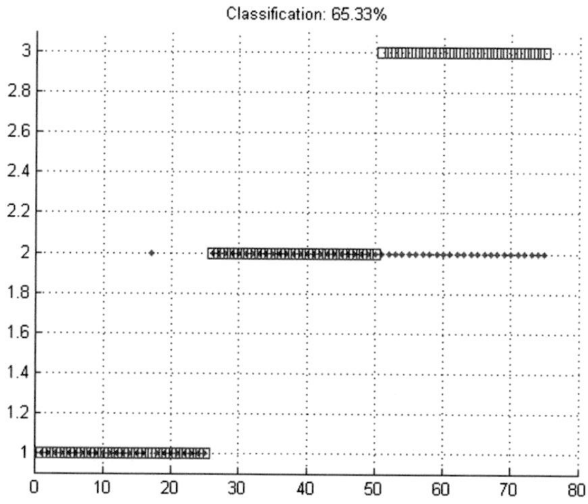

Fig. 8 Classification results for Iris data set using Mamdani T2 FIS designed by IT2FCM

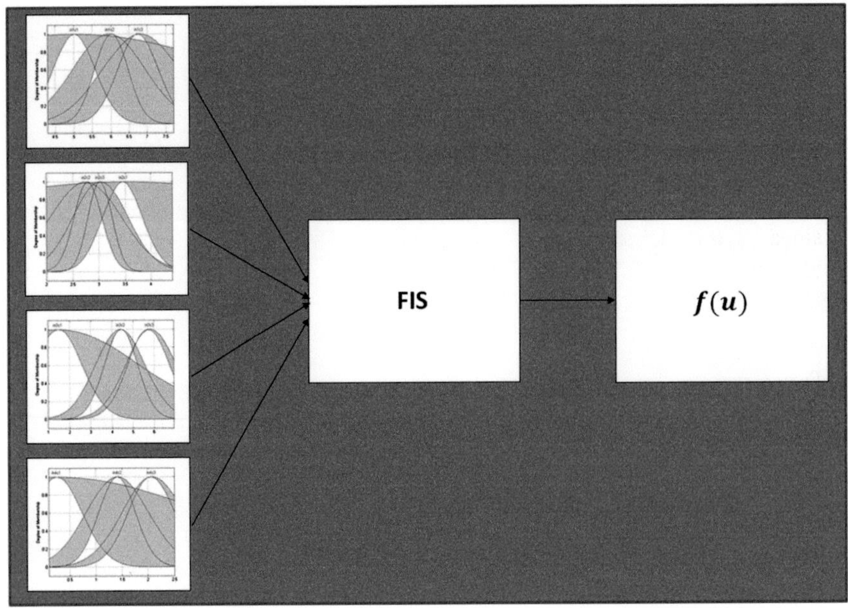

Fig. 9 Sugeno T2 FIS designed by IT2FCM for classification of the Iris data set

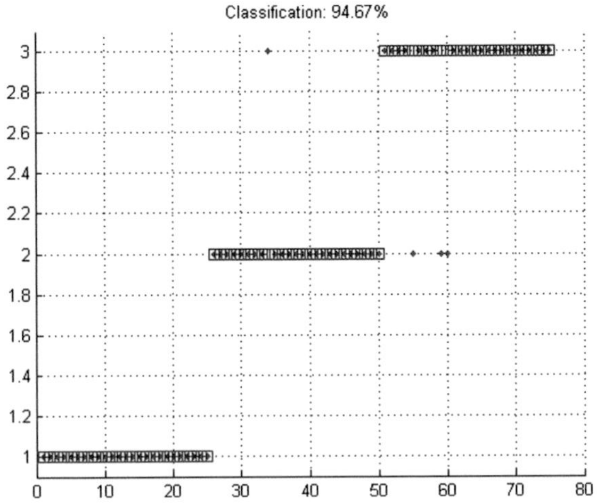

Fig. 10 Classification results for Iris data set using the sugenot2 FIS designed by IT2FCM

5 Conclusions

The IT2FCM algorithm allows us to design FIS from clusters found in a data set, allowing us to create FIS for classification or time series prediction problems from clusters found in input and output data, given to the algorithm.

The results obtained by the Sugeno FIS designed by the IT2FCM algorithm for the problem of classification of iris flower was good achieving a 94.67 % of correct classification, but the result obtained by Mamdani FIS designed by IT2FCM algorithm for the above problem was not as good as expected getting a 65.33 % of correct classification.

The results obtained by the Sugeno FIS designed by the IT2FCM algorithm for the problem of time series prediction was 0.1133 of error prediction, but the results obtained by the Mamdani FIS designed by the IT2FCM algorithm for the above problem was 0.1646 of error prediction.

References

1. Bezdek, J.: Pattern Recognition with Fuzzy Objective Function Algorithms. Plenum, New York (1981)
2. Ceylan, R., Özbay, Y., Karlik, B.: "A novel approach for classification of ECG arrhythmias: Type-2 fuzzy clustering neural network. Expert Syst. Appl. **36**(3), 6721–6726 (2009)
3. Chang, X., Li, W., Farrell, J.: A C-means clustering based fuzzy modeling method. In: The Ninth IEEE International Conference on Fuzzy Systems, 2000. FUZZ IEEE 2000, vol. 2, pp. 937–940 (2000)
4. Choi, B., Rhee, F.: Interval type-2 fuzzy membership function generation methods for pattern recognition. Inf. Sci. **179**(13), 2102–2122 (2009)
5. Gustafson, D.E., Kessel, W.C.: Fuzzy clustering with a fuzzy covariance matrix. In: Proceeding of IEEE Conference on Decision Control, pp. 761–766. San Diego, CA (1979)
6. Hirota, K., Pedrycz, W.: Fuzzy computing for data mining. Proc. IEEE **87**(9), 1575–1600 (1999)
7. Hwang, C., Rhee, F.: Uncertain fuzzy clustering: interval type-2 fuzzy approach to C-means. IEEE Trans. Fuzzy Syst. **15**(1), 107–120 (2007)
8. Iyer, N.S., Kendel, A., Schneider, M.: Feature-based fuzzy classification for interpretation of mamograms. Fuzzy Sets Syst. **114**, 271–280 (2000)
9. Jang, J.S.R.: ANFIS: Adaptive-network-based fuzzy inference systems. IEEE Trans. on Syst. Man Cybern. **23**, 665–685 (1992)
10. Karnik, N., Mendel, M.: Operations on type-2 set. Fuzzy Set Syst. **122**, 327–348 (2001)
11. Krishnapuram, R., Keller, J.: A possibilistic approach to clustering. IEEE Trans. Fuzzy Syst. **1** (2), 98–110 (1993)
12. Kruse, R., Döring, C., Lesot, M.J.: Fundamentals of fuzzy clustering. In: Advances in Fuzzy Clustering and its Applications; Wiley, The Atrium, Southern Gate, Chichester, West Sussex PO19 8SQ, England, pp. 3–30 (2007)
13. Melin, P., Soto, J., Castillo, O., Soria, J.: A new approach for time series prediction using ensembles of ANFIS models. Experts Syst. Appl. El-Sevier **39**(3), 3494–3506 (2012)
14. Mendel, J.: Uncertain Rule-Based Fuzzy Logic Systems: Introduction and new directions, pp. 213–231. Prentice-Hall, Inc., Upper-Saddle River (2001)
15. Pal, N.R., Pal, K., Keller, J.M., Bezdek, J.C.: A possibilistic fuzzy c-means clustering algorithm. IEEE Trans. Fuzzy Syst. **13**(4), 517–530 (2005)
16. Philips, W.E., Velthuinzen, R.P., Phuphanich, S., Hall, L.O., Clark, L.P., Sibiger, M.L.: Application of fuzzy c-means segmentation technique for tissue differentation in MR images of hemorrhagic gliobastoma multifrome. Magn. Reson. Imaging **13**(2), 277–290 (1995)
17. Pulido, M., Mancilla, A., Melin, P.: Ensemble neural networks with fuzzy logic integration for complex time series prediction. IJIEI **1**(1), 89–103 (2010)

18. Pulido, M., Mancilla, A., Melin, P.: An ensemble neural network architecture with fuzzy response integration for complex time series prediction. In: Evolutionary Design of Intelligent Systems in Modeling, Simulation and Control, pp. 85–110 (2009)
19. Pulido, M., Mancilla, A., Melin, P.: Ensemble neural networks with fuzzy integration for complex time series prediction. In: Bio-inspired Hybrid Intelligent Systems for Image Analysis and Pattern Recognition, pp. 143–155 (2009)
20. Rubio, E.; Castillo, O.: Interval type-2 fuzzy clustering for membership function generation. In: 2013 IEEE Workshop on Hybrid Intelligent Models and Applications (HIMA), pp. 13–18, 16–19 April 2013
21. Rubio, E.; Castillo, O.: Optimization of the interval type-2 fuzzy C-means using particle swarm optimization. In: Proceedings of NABIC 2013, pp. 10–15. Fargo, USA (2013)
22. Soto J., Castillo O., Soria J.: A New approach for time series prediction using ensembles of ANFIS models. In: Soft Computing for Intelligent Control and Mobile Robotics, Springer, vol. 318, pp. 483 (2015)
23. Soto J., Melin P., Castillo O.: Time series prediction using ensembles of neuro-fuzzy models with interval type-2 and type-1 fuzzy integrators. In: Proceedings of International Joint Conference on Neural Nerworks, Dallas, pp. 189–194 Texas, USA, 4–9 Aug 2013
24. Wang C., Zhang J.P.: Time series prediction based on ensemble ANFIS. In: Proceedings of the Fourth International Conference on Machine Learning and Cybernetics, Guangzhou, 18–21 Aug 2005
25. Yang, M.-S., Hu, Y.-J., Lin, K.C.-R., Lin, C.C.-L.: Segmentation techniques for tissue differentiation in MRI of ophthalmology using fuzzy clustering algorithms. Magn. Reson. Imaging **20**, 173–179 (2002)
26. Yen, J., Langari, R.: Fuzzy Logic: Intelligence, Control, and Information. Prentice Hall, Upper Saddle River (1999)
27. Zadeh, L.A.: The concept of a linguistic variable and its application to approximate reasoning-I. Inform. Sci. **8**(3), 199–249 (1975)

Multiple-Criteria Evaluation in the Fuzzy Environment Using the FuzzME Software

Pavel Holeček, Jana Talašová and Jan Stoklasa

Abstract This chapter describes a software tool for fuzzy multiple-criteria evaluation called FuzzME. The chapter will show the reader in an easy-to-read style how to apply the software for solving a broad range of fuzzy MCDM problems. The mathematical foundation on which the FuzzME software is built will be described and demonstrated on an example. The FuzzME implements a complete system of fuzzy methods. A common feature of all these methods is the type of evaluation that is well-suited to the paradigm of fuzzy set theory. All evaluations in the presented models are in the form of fuzzy numbers expressing the extent to which goals of evaluation have been fulfilled. The system of fuzzy methods can deal with different types of interaction among criteria of evaluation. If there is no interaction among criteria, then either fuzzy weighted average, fuzzy OWA operator, or fuzzified WOWA operator is used to aggregate partial evaluations (depending on evaluator's requirements on the type of evaluation). If interactions among criteria are in the form of redundancy or complementarity, then fuzzified discrete Choquet integral is an appropriate aggregation operator. In case of more complex interactions, the aggregation function is described by an expertly defined base of fuzzy rules. The FuzzME also contains additional tools which make it possible to perform analysis of the designed evaluation model and to adjust it easily.

P. Holeček (✉) · J. Talašová · J. Stoklasa
Faculty of Science, Department of Mathematical Analysis and Applications of Mathematics, Palacký University in Olomouc, Olomouc, Czech Republic
e-mail: pavel.holecek@upol.cz
URL: http://fuzzymcdm.upol.cz

J. Talašová
e-mail: jana.talasova@upol.cz

J. Stoklasa
e-mail: jan.stoklasa@upol.cz

M. Collan et al. (eds.), *Fuzzy Technology*, Studies in Fuzziness and Soft Computing 335, DOI 10.1007/978-3-319-26986-3_9

1 Introduction

Multiple-criteria evaluation problems are very common in the practice. For example a bank needs to evaluate the credibility of a company in order to decide if it should be granted a credit or not; or a company hiring new employees has to evaluate all the candidates for the job to determine who of them is the best one. Because the values of the qualitative criteria are given expertly, they contain some uncertainty. Moreover, the values of the quantitative criteria are not always known precisely, too. Therefore, it is reasonable to use fuzzy models of multiple-criteria evaluation, which can take this uncertainty into account.

There are plenty of MCDM software tools. However, only few of them employ fuzzy sets. One of them, the FuzzME software, will be presented in this chapter. The mathematical methods used in the software will be summarized and the possibilities of the software tool will be demonstrated on an example.

2 Example

The theory will be shown on the following simplified example, which is designed so that the possibilities of the FuzzME software could be demonstrated easily. Later, we will mention three real applications.

Let us assume that a software company wants to hire a new programmer. Many candidates applied for this vacancy. The company has to evaluate the candidates and choose the best of them. The HR-manager has listed the following criteria that have importance for the company: *language skills*, *C# programming language knowledge*, *Java programming language knowledge*, *ability to work in teams*, *motivation*, *length of practice in years* and the *references from the previous employers*.

The given problem contains many obstacles: (1) there are both qualitative and quantitative criteria, (2) some of the criteria values can be very uncertain (for example the motivation is evaluated just by the impression of the candidate on the HR-manager), (3) some criteria values could not be known at all, and, finally, (4) the evaluation criteria are not fully independent. We will show that all of these difficulties can be overcome with the presented model.

3 The Used Evaluation Model

3.1 The Basic Notions of the Fuzzy Set Theory

In this section, we will mention briefly some of the notions that will be used throughout this chapter. More detailed information on the fuzzy set theory can be found for example in [4].

A fuzzy set A on a universal set X is characterized by its membership function $A : X \to [0, 1]$. *Ker A* denotes a kernel of A, $Ker\, A = \{x \in X \mid A(x) = 1\}$. For any $\alpha \in [0, 1]$, A_α denotes an α-cut of A, $A_\alpha = \{x \in X \mid A(x) \geq \alpha\}$. A support of A is defined as $Supp\, A = \{x \in X \mid A(x) > 0\}$. The height of the fuzzy set A, *hgt A*, is defined as $hgt\, A = \sup\{A(x) \mid x \in X\}$.

The symbol $\tilde{0}$ will denote the crisp 0 in form of a fuzzy singleton, i.e. $\tilde{0}(0) = 1$ and $\tilde{0}(x) = 0$, for $x \in \mathfrak{R}$, $x \neq 0$. Similarly, the fuzzy set $\tilde{1}$ will be a fuzzy singleton representing the crisp 1, i.e. $\tilde{1}(1) = 1$ and $\tilde{1}(x) = 0$, for $x \in \mathfrak{R}, x \neq 1$.

A fuzzy number is a fuzzy set C on the set of all real numbers $\mathfrak{R}$ which satisfies the following conditions: (a) the kernel of C, *Ker C*, is not empty, (b) the α-cuts of C, C_α, are closed intervals for all $\alpha \in (0, 1]$, (c) the support of C, *Supp C*, is bounded.

Real numbers $c^1 \leq c^2 \leq c^3 \leq c^4$ are called significant values of the fuzzy number C if the following holds: $[c^1, c^4] = Cl(Supp\ C)$, $[c^2, c^3] = Ker\ C$, where $Cl(Supp\ C)$ denotes a closure of *Supp C*. A fuzzy number C is said to be linear, if its membership function between each pair of the neighboring significant values is linear. The linear fuzzy number C is called triangular if $c^2 = c^3$, otherwise it is called trapezoidal. In the examples, linear fuzzy numbers will be described by their significant values. Therefore, we will write such a fuzzy number as $C = (c^1, c^2, c^3, c^4)$ if C is trapezoidal, or simply $C = (c^1, c^2, c^4)$ if C is a triangular fuzzy number.

Any fuzzy numbers C can be described, beside its membership function, in an alternative way by a pair of functions $\underline{c} : [0, 1] \to \mathfrak{R}, \overline{c} : [0, 1] \to \mathfrak{R}$ defined as follows

$$
\begin{aligned}
C_\alpha &= [\underline{c}(\alpha), \overline{c}(\alpha)] \quad \text{for all } \alpha \in (0, 1], \text{ and} && (1)\\
Cl(Supp\ C) &= [\underline{c}(0), \overline{c}(0)]. && (2)
\end{aligned}
$$

Then, the fuzzy number C can be written in the form $C = \left\{ \left[\underline{c}(\alpha), \quad \overline{c}(\alpha) \right], \quad \alpha \in [0, 1] \right\}$.

Let A be a fuzzy number, $A = \{[\underline{a}(\alpha), \overline{a}(\alpha)], \alpha \in [0, 1]\}$. Let a constant $c \in \mathfrak{R}$ be given. Multiplying the fuzzy number A by the real number c we obtain a fuzzy number $c \cdot A$ defined as follows

$$
c \cdot A = \begin{cases} \left\{ \left[c \cdot \underline{a}(\alpha), \quad c \cdot \overline{a}(\alpha) \right], \quad \alpha \in [0, 1] \right\} \text{ for } c \geq 0 \\ \left\{ \left[c \cdot \overline{a}(\alpha), \quad c \cdot \underline{a}(\alpha) \right], \quad \alpha \in [0, 1] \right\} \text{ for } c < 0. \end{cases} \tag{3}
$$

Let A and B be fuzzy numbers, $A = \{[\underline{a}(\alpha), \overline{a}(\alpha)], \alpha \in [0, 1]\}$, $B = \{[\underline{b}(\alpha), \overline{b}(\alpha)], \alpha \in [0, 1]\}$. Their sum, $A + B$, is a fuzzy numbers given as follows:

$$
A + B = \left\{ \left[\underline{a}(\alpha) + \underline{b}(\alpha), \quad \overline{a}(\alpha) + \underline{b}(\alpha) \right], \quad \alpha \in [0, 1] \right\} \tag{4}
$$

Similarly, a subtraction, multiplication and division of fuzzy numbers could be defined (see e.g. [4, 8]). They, however, will not be needed in the further text.

A linguistic variable [15] is defined as a quintuple $(\mathcal{V}, \mathcal{T}(\mathcal{V}), X, G, M)$, where $\mathcal{V}$ is a name of the variable, $\mathcal{T}(\mathcal{V})$ is a set of its linguistic values, X is a universal set on which the meanings of the linguistic values are defined, G is a syntactic rule for generating values in $\mathcal{T}(\mathcal{V})$, and M is a semantic rule which maps each linguistic value $\mathcal{C} \in \mathcal{T}(\mathcal{V})$ to its mathematical meaning, $C = M(\mathcal{C})$, which is a fuzzy set on X.

An ordering of fuzzy numbers is defined as follows: a fuzzy number C is greater than or equal to a fuzzy number D, $C \geq D$, if $C_\alpha \geq D_\alpha$ for all $\alpha \in (0, 1]$. The inequality of the α-cuts $C_\alpha \geq D_\alpha$ is the inequality of intervals $C_\alpha = [\underline{c}(\alpha), \overline{c}(\alpha)]$, $D_\alpha = [\underline{d}(\alpha), \overline{d}(\alpha)]$ which is defined as

$$[\underline{c}(\alpha), \overline{c}(\alpha)] \geq [\underline{d}(\alpha), \overline{d}(\alpha)] \text{ if, and only if, } \underline{c}(\alpha) \geq \underline{d}(\alpha) \text{ and } \overline{c}(\alpha) \geq \overline{d}(\alpha).$$

This relation is only a partial ordering and many fuzzy numbers can be incomparable this way. However, it is possible to order any fuzzy numbers according to their centers of gravity [4].

3.2 *The Basic Structure of the Evaluation*

The basic structure used for the evaluation is called a goals tree. The root of the tree represents the main goal. The main goal is consecutively divided into partial goals of lower level. The partial goals at the end of the branches are connected with either qualitative or quantitative criteria.

All evaluations in the presented models are in the form of fuzzy numbers on the interval [0, 1]. They express how much have the particular partial goals been fulfilled. The evaluation $\tilde{0}$ means that the particular alternative does not fulfill our goal at all, while the evaluation $\tilde{1}$ means that we are fully satisfied with the alternative with respect to the particular partial goal.

When an alternative is evaluated, the evaluations with respect to the criteria at the end of the goals tree branches are calculated first. Those evaluations are, as it has already been said, fuzzy numbers on [0, 1]. These evaluations are then aggregated together to obtain the evaluation of the partial goal on the higher level in the goals tree. For the aggregation, we can use one of the supported aggregation operators (fuzzy weighted average, fuzzy OWA operator, fuzzified WOWA, or fuzzy Choquet integral) or a fuzzy expert system. The process of aggregation is repeated until the root of the goals tree is reached. The evaluation in the root of the tree is the overall evaluation of the alternative.

The resulting evaluations are fuzzy numbers. The FuzzME software presents the result in several forms and thus ensuring that they will be easy to interpret for the decision-maker. The results are provided not only in the numerical (e.g. center of gravity, kernel) form, but they are also described verbally by a linguistic approximation (for more information see [8]) and they are presented in a graphical form. The alternatives can be also ordered by the centers of gravity of their evaluations.

The goals tree for our example is depicted in the Fig. 1. The next sections will describe each part of the evaluation process in more detail.

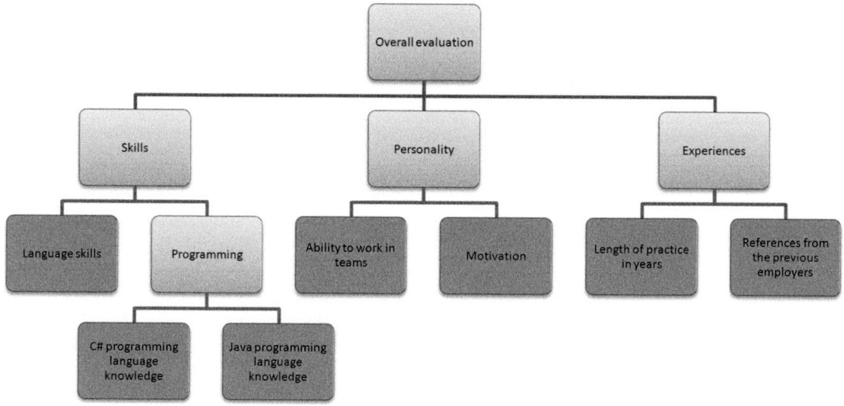

Fig. 1 A goals tree for evaluation of the candidates by the company

3.3 Evaluation Criteria

Two types of criteria can be used—qualitative and quantitative ones. Each of them is evaluated in a different way.

For qualitative criteria, a linguistic variable has to be defined in advance. When an alternative is evaluated according to a qualitative criterion, the decision-maker selects the value from the linguistic variable, which is the best-fitting for the alternative. Let us assume, the linguistic variable from the Fig. 2 is used for all qualitative criteria in our example. *Language skills* are an example of such a criterion. When a candidate is evaluated, the expert chooses the best fitting term for his/her language skills. The expert can asses them for example to be *very good*. As an evaluation of the alternative (candidate) according to this criterion, the fuzzy number that models meaning of the term *very good* will be taken, i.e. the trapezoidal fuzzy number with the significant values $(0.78, 0.89, 1, 1)$.

For quantitative criteria, an evaluating function has to be defined first. An example of a quantitative criterion is the length of practice of the candidate measured in years.

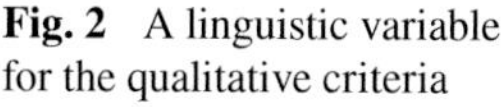
Fig. 2 A linguistic variable for the qualitative criteria

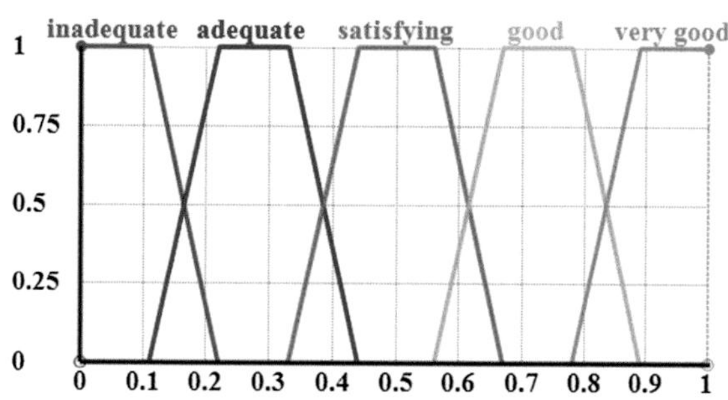

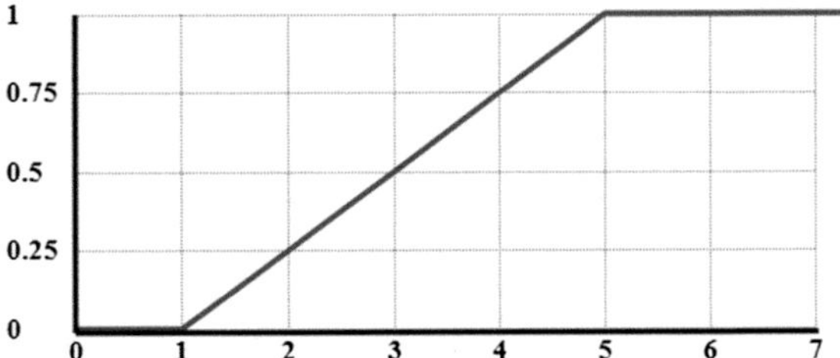

Fig. 3 An evaluating function for *Length of practice* (in years)

The evaluation function for this criterion is depicted in the Fig. 3. By this function, the expert expresses that the candidates, who do not have more than one year of practice, are completely unsatisfactory, whereas the candidates with 5 or more years of practice are fully satisfactory for the company. For example, a candidate with 3 years of practice fulfills the goal in the degree 0.5. If the number of year is not known precisely, it could be given by a fuzzy number and the resulting evaluation is then calculated according to the extension principle. For example, if a candidate has 1 year of experiences with the software and methods used in the company, another year of experience with the a similar software and methods and, finally, two years, when he/she worked with something completely different, which can however turn out to be helpful, we could model the length of practice by a triangular fuzzy number with the significant values $(2, 3, 4)$. Using the extension principle, we obtain the fuzzy evaluation of this criterion as a triangular fuzzy number $(0.25, 0.5, 0.75)$.

3.4 Aggregation

The evaluations of the partial goals on the lower level are aggregated together so that the evaluation of the partial goal on the higher level would be obtained. For this, the decision-maker can employ multiple methods. Most of them are fuzzified versions of well-known aggregation operators. Because of the space limitation, we will not be able to describe all of them in detail. For their full description, please see [8].

It is possible to use different aggregation methods in the same goals tree. Moreover, the FuzzME contains algorithms to make a subsequent change of one aggregation method to another one really easy. If the expert used the fuzzy weighted average originally and then he/she would like to change the method to, for example, the fuzzified Choquet integral, the corresponding FNV-fuzzy measure is generated in the FuzzME automatically. The detailed description how is this done can be found in [7].

The choice of the aggregation method depends on the relationship among the partial goals and on the requirements of the decision-maker on the behavior of the evaluating function.

We will show all of the supported methods on the partial evaluation of *Programming*, which is obtained by the aggregation of the candidate's knowledge evaluations of *C#* and Java programming languages.

3.5 Fuzzy Weighted Average

The weighted average is probably the most commonly used aggregation operator. One of its fuzzified versions, which is used in the FuzzME, has been proposed in [10]. It uses the structure of normalized fuzzy weights, which represent an uncertain division of the whole into parts.

Definition 1 Fuzzy numbers $W_1, \ldots, W_m$ defined on $[0, 1]$ form normalized fuzzy weights [10] if for any $i \in \{1, \ldots, m\}$ and any $\alpha \in (0, 1]$ it holds that for any $w_i \in W_{i\alpha}$ there exist $w_j \in W_{j\alpha}$, $j = 1, \ldots, m$, $j \neq i$, such that

$$w_i + \sum_{j=1, j\neq i}^{m} w_j = 1. \tag{5}$$

Definition 2 The fuzzy weighted average of the partial fuzzy evaluations, i.e., of fuzzy numbers $U_1, \ldots, U_m$ defined on $[0, 1]$, with the normalized fuzzy weights $W_1, \ldots, W_m$ is a fuzzy number U on $[0, 1]$ whose membership function is defined for any $u \in [0, 1]$ as follows

$$\begin{aligned} U(u) = \max\{\min\{W_1(w_1), \ldots, W_m(w_m), U_1(u_1), \ldots, U_m(u_m)\} \\ \mid \sum_{i=1}^{m} w_i \cdot u_i = u, \ \sum_{i=1}^{m} w_i = 1, \ w_i, u_i \in [0, 1], i = 1, \ldots, m\}. \end{aligned} \tag{6}$$

If the company from our example prefers the candidates to have knowledge of Java, they could use the fuzzy weighted average with the following normalized fuzzy weights for both of the criteria: $W_{C\#} = (0.2, 0.3, 0.4)$, $W_{Java} = (0.6, 0.7, 0.8)$.

Figure 4 shows the graph of the evaluation function. Because the result is, in our case, a triangular fuzzy number, three surfaces are plotted in the graph. Each of them represents one significant value of the resulting fuzzy number. This way, we are able to visualize the entire information about the result, and not just some of its characteristics (such as the center of gravity). This will help to understand the behavior of the evaluation function better. On the x-axis, there is an evaluation of the C# knowledge, on the y-axis, there is an evaluation of the Java knowledge. In order to be able to construct the graph, we assume only crisp values of those two partial evaluations (i.e. fuzzy singletons on [0, 1]). For the comparison, Fig. 4a shows the result with the crisp weights $W_{C\#} = 0.3$, $W_{Java} = 0.7$.

When we study the Fig. 4, we can see that if both of the partial evaluations are equal, the result is a fuzzy singleton. The more the two evaluations differ, the more

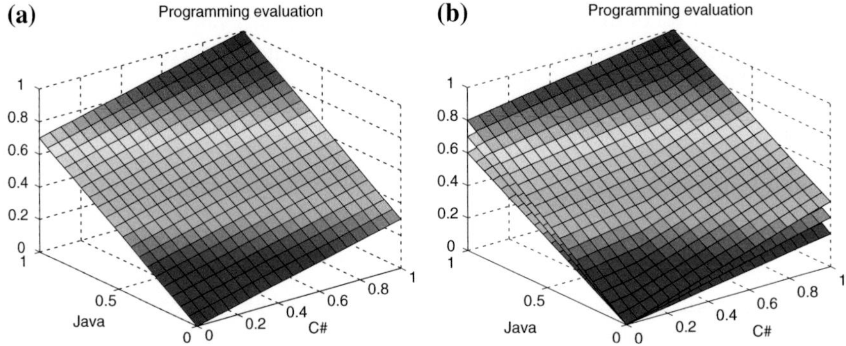

Fig. 4 Evaluation by fuzzy weighted average. **a** In the crisp case. **b** In the fuzzy case

uncertainty is contained in the overall evaluation. This is an important property of the fuzzy weighted average, which is very beneficial for the multiple-criteria decision-making.

To make this behavior clearer, let us consider a fuzzy weighted average of 4 partial evaluations U_1, U_2, U_3, U_4 with uniform normalized fuzzy weights ($W_1 = W_2 = W_3 = W_4 = (0.05, 0.25, 0.45)$). The partial evaluations can represent evaluations of various aspects of a bank client. Let us consider the situation when the client evaluation is average according to all of the four aspects and another situations, when the client is evaluated as excellent according the half of the aspects and completely unsatisfactory according to the rest of them. Then evaluation with the regular weighted average using uniform non-fuzzy weights average would make no difference between those two cases and evaluate the client as average (0.5). If the fuzzy weighted average is used the two fuzzy evaluations will have the same center of gravity, however, the latter one will be much more uncertain. The fuzzy weighted average takes into consideration also the dispersion of the aggregated values. This is shown in the Fig. 5a, b. In the Fig. 5a, the aggregated values are closer to each other and therefore the result is less uncertain. On the other hand, in Fig. 5b, the aggregated values differ more so the resulting fuzzy evaluation is more uncertain.

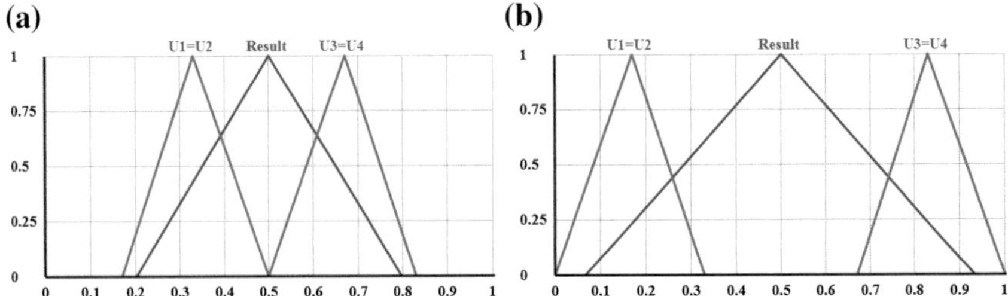

Fig. 5 Comparison of results of the fuzzy weighted average when the aggregated values are **a** close to each other, **b** further apart

3.6 Fuzzy OWA Operator

Sometimes it is not necessary to set the importances directly for the individual partial goals (programming languages in this case). Instead, the importances should be set according to the order of the evaluation (in which programming language is the programmer more skilled). This can be done by the OWA operator [14]. The FuzzME implements its fuzzified version that has been introduced in [12]. Similarly to the fuzzy weighted average mentioned before, it also uses the structure of normalized fuzzy weights.

Definition 3 The fuzzy OWA of the partial fuzzy evaluations, i.e., of fuzzy numbers $U_1, \ldots, U_m$ defined on $[0, 1]$, with normalized fuzzy weights $W_1, \ldots, W_m$ is a fuzzy number U on $[0, 1]$ whose membership function is defined for any $u \in [0, 1]$ as follows

$$U(u) = \max\{\min\{W_1(w_1), \ldots, W_m(w_m), U_1(u_1), \ldots, U_m(u_m)\} \mid \sum_{i=1}^{m} w_i \cdot u_{\phi(i)} = u, \sum_{i=1}^{m} w_i = 1, w_i, u_i \in [0, 1], i = 1, \ldots, m\}, \tag{7}$$

where ϕ denotes such a permutation of the set of indices $\{1, \ldots, m\}$ that $u_{\phi(1)} \geq u_{\phi(2)} \geq \ldots \geq u_{\phi(m)}$.

If the company prefers neither of the programming languages but wants the candidate to be good at least in one of them, a fuzzy OWA operator with the following normalized fuzzy weights can be used: $W_1 = (0.7, 0.8, 0.9)$, $W_2 = (0.1, 0.2, 0.3)$. This way, the higher importance will be attached to the programming language that the candidate knows better. However, the evaluation of knowledge of the other programming language will be taken into account, too. The behavior of the evaluation function can be seen in the Fig. 6b. It can be compared with the crisp OWA operator (using weights $W_1 = 0.8$, $W_2 = 0.2$) whose graph is depicted in the Fig. 6a. It can be

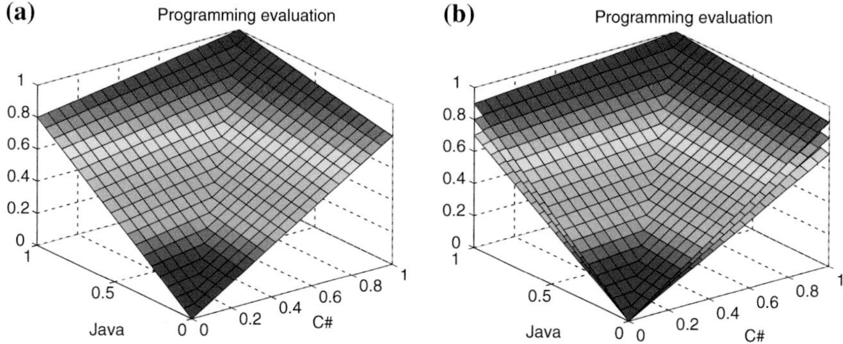

Fig. 6 Evaluation by OWA operator. **a** In the crisp case. **b** In the fuzzy case

seen that the behavior when normalized fuzzy weights are used is similar as in case of the fuzzy weighted average. The more dispersion between the aggregated partial evaluations, the more uncertain result.

3.7 Fuzzifed WOWA Operator

The WOWA operator, which combines advantages of both weighted average and the OWA, has been proposed by Torra in [13]. In this case, two m-tuples of weights are used. The first one, $(w_1, \ldots, w_m)$, corresponds to the individual partial goals (as in case of the weighted average), the latter one, $(p_1, \ldots, p_m)$, is connected with the order of the evaluation (as in OWA operator).

The FuzzME uses the fuzzified WOWA operator defined in [8]. In this case, the input variables are fuzzy numbers but the weights are crisp. Otherwise, all the other aggregation methods in FuzzME accept both the input variables and the importances expressed in the form of fuzzy numbers. The implementation of the fuzzified WOWA with fuzzy weights will be the topic of the future research.

Definition 4 Let U_i, $i = 1, \ldots, m$, be fuzzy numbers and let w_i and p_i, $i = 1, \ldots, m$, be two set of normalized (real) weights. Then the result of the aggregation by a fuzzified WOWA operator is a fuzzy number $U = \{[\underline{u}(\alpha), \overline{u}(\alpha)], \alpha \in [0, 1]\}$ defined for any $\alpha \in [0, 1]$ as follows

$$\underline{u}(\alpha) = \sum_{i=1}^{m} \omega_i^L \cdot \underline{u}_{\sigma(i)}(\alpha), \tag{8}$$

$$\overline{u}(\alpha) = \sum_{i=1}^{m} \omega_i^R \cdot \overline{u}_{\chi(i)}(\alpha), \tag{9}$$

where σ, and χ are permutations of the set of indices $\{1, \ldots, m\}$ such that $\underline{u}_{\sigma(1)}(\alpha) \geq \cdots \geq \underline{u}_{\sigma(m)}(\alpha)$ and $\overline{u}_{\chi(1)}(\alpha) \geq \cdots \geq \overline{u}_{\chi(m)}(\alpha)$. The weights ω_i^L and ω_i^R, $i = 1, \ldots, m$, are defined for the given α as

$$\omega_i^L = z(\sum_{j \leq i} w_{\sigma(j)}) - z(\sum_{j < i} w_{\sigma(j)}), \tag{10}$$

$$\omega_i^R = z(\sum_{j \leq i} w_{\chi(j)}) - z(\sum_{j < i} w_{\chi(j)}), \tag{11}$$

where z is a nondecreasing piece-wise linear function interpolating the following points

$$\{(0, 0)\} \cup \{(i/m, \sum_{j \leq i} p_j)\}_{i=1,\ldots,m}. \tag{12}$$

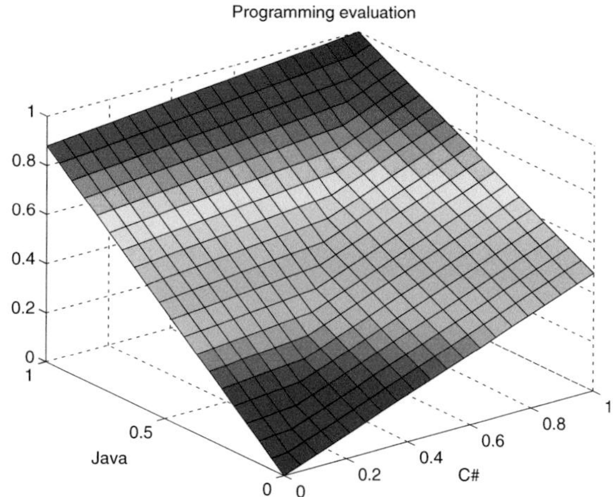

Fig. 7 Evaluation by fuzzified WOWA operator (only version with crisp weights is implemented in FuzzME)

As it has been mentioned, the fuzzified WOWA requires two set of normalized (real) weights—one, which is connected to the individual partial goals as in case of the weighted average, and another one, which is connected to the order of their evaluations as in the case of OWA. The company could therefore use the following sets of weights to combine the previous two approaches: $w_{C\#} = 0.3$, $w_{Java} = 0.7$, and $p_1 = 0.8, p_2 = 0.2$.

The evaluating function is depicted in the Fig. 7. As this version of fuzzified WOWA uses crisp weights only (and therefore, if evaluations of the C# and Java are crisp, the result will be also crisp), the graph coincides with the graph of the crisp WOWA.

3.8 Fuzzified Choquet Integral

The aggregation methods mentioned so far assumed that there are no interactions among the aggregated partial goals. Otherwise, their result could be misleading. If the interactions are present and they have the character of redundancy or complementarity, the discrete Choquet integral [3] can be used for the evaluation.

The FuzzME provides the possibility to aggregate the partial evaluations by the fuzzified Choquet integral that has been introduced in [2]. This fuzzification uses a FNV-fuzzy measure (whose values are fuzzy numbers) instead of a fuzzy measure (whose values are, despite its name, crisp numbers). The FNV-fuzzy measure and the fuzzified Choquet integral are defined as follows.

Definition 5 Let $G = \{G_1, \ldots, G_m\}$ be a nonempty finite set, $\wp(G)$ be the family of all its subsets. Then, a FNV-fuzzy measure on G is a set function $\tilde{\mu} : \wp(G) \rightarrow \mathcal{F}_N([0, 1])$ satisfying the following conditions:

- $\tilde{\mu}(\emptyset) = \tilde{0}$, $\tilde{\mu}(G) = \tilde{1}$, and
- $C \subseteq D$ implies $\tilde{\mu}(C) \leq \tilde{\mu}(D)$ for any $C, D \in \wp(G)$.

Definition 6 The fuzzified Choquet integral of a FNV-function F with respect to the FNV-fuzzy measure $\tilde{\mu}$ is defined as a fuzzy number U with a membership function given for any $u \in [0, 1]$ as

$$U(u) = \max \Big\{ \min\{U_1(u_1), \dots, U_m(u_m), \tilde{\mu}(B_{\rho(1)})(\mu_1), \dots, \tilde{\mu}(B_{\rho(m)})(\mu_m)\} \mid \quad (13)$$
$$u = (C)\int_G f d\mu, \text{ where } f : G \to [0,1] \text{ such that } f(G_i) = u_i, i = 1, \dots, m,$$
$$\mu \text{ is a fuzzy measure on } G \text{ such that } \mu(B_{\rho(i)}) = \mu_i, i = 1, \dots, m \Big\},$$

where ρ denotes such a permutation of the set of indices $\{1, \dots, m\}$ that $u_{\rho(1)} \leq u_{\rho(2)} \leq \dots \leq u_{\rho(m)}$ and $B_{\rho(i)} = \{G_{\rho(i)}, \dots, G_{\rho(m)}\}$. By definition, we will set $B_{\rho(m+1)} = \emptyset$.

The C# and Java programming languages have lots in common. If a programmer excels in one of them, he/she can learn the other one quite quickly. We can see that there is a relationship of redundancy among these criteria and the Choquet integral is ideal for these cases. In the FuzzME, its fuzzified version is available, where not only the aggregated values, but also the measure values are represented by fuzzy numbers. The following FNV-fuzzy measure can be used for this case: $\tilde{\mu}(\emptyset) = \tilde{0}$, $\tilde{\mu}(\text{C\#}) = (0.4, 0.5, 0.6)$, $\tilde{\mu}(\text{Java}) = (0.7, 0.8, 0.9)$, $\tilde{\mu}(\text{C\#, Java}) = \tilde{1}$.

The graph of the resulting evaluation function can be seen in the Fig. 8b. For comparison, the graph of the crisp Choquet integral is in the Fig. 8a (the used fuzzy measure is $\mu(\emptyset) = 0$, $\mu(\text{C\#}) = 0.5$, $\mu(\text{Java}) = 0.8$, $\mu(\text{C\#, Java}) = 1$).

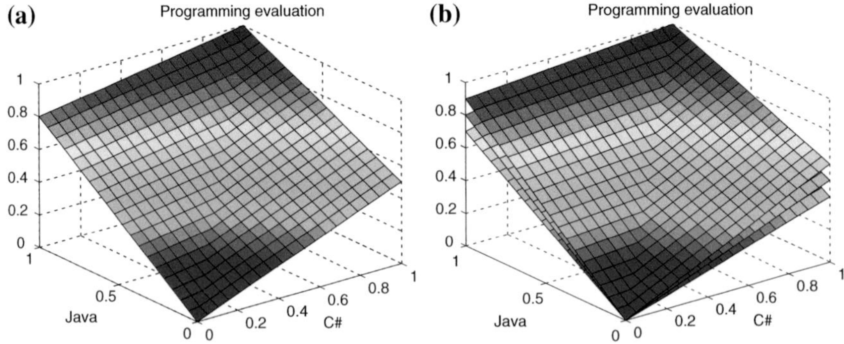

Fig. 8 Evaluation by the Choquet integral. **a** The crisp case (a fuzzy measure is used) **b** The fuzzy case (a FNV-fuzzy measure is used)

3.9 Fuzzy Expert System

If there are complex relationships among the aggregated partial goals that cannot be handled even by the fuzzified Choquet integral, a fuzzy expert system can be always used. The evaluation function is expressed in form of the fuzzy rule base, which has the following form.

$$\begin{aligned}
&\text{If } \mathcal{E}_1 \text{ is } \mathcal{U}_{1,1} \text{ and } \dots \text{ and } \mathcal{E}_m \text{ is } \mathcal{U}_{1,m}, \text{then } \mathcal{E} \text{ is } \mathcal{U}_1,\\
&\text{If } \mathcal{E}_1 \text{ is } \mathcal{U}_{2,1} \text{ and } \dots \text{ and } \mathcal{E}_m \text{ is } \mathcal{U}_{2,m}, \text{then } \mathcal{E} \text{ is } \mathcal{U}_2,\\
&\dots\dots\dots\dots\dots\dots\dots\dots\dots\dots\dots\dots\dots\dots\dots\dots\dots\\
&\text{If } \mathcal{E}_1 \text{ is } \mathcal{U}_{n,1} \text{ and } \dots \text{ and } \mathcal{E}_m \text{ is } \mathcal{U}_{n,m}, \text{then } \mathcal{E} \text{ is } \mathcal{U}_n,
\end{aligned}$$

where for $i = 1, 2, \dots, n, j = 1, 2, \dots, m$:

- $(\mathcal{E}_j, \mathcal{T}(\mathcal{E}_j), [0, 1], M_j, G_j)$ are linguistic variables representing partial evaluations,
- $\mathcal{U}_{ij} \in \mathcal{T}(\mathcal{E}_j)$ are their linguistic values and $U_{ij} = M_j(\mathcal{U}_{ij})$ are fuzzy numbers on $[0, 1]$ representing their meanings,
- $(\mathcal{E}, \mathcal{T}(\mathcal{E}), [0, 1], M, G)$ is a linguistic variable representing the overall evaluation,
- $\mathcal{U}_i \in \mathcal{T}(\mathcal{E})$ are its linguistic values and $U_i = M(\mathcal{U}_i)$ are fuzzy numbers on $[0, 1]$ representing their meanings.

Then, one of many inference algorithms can be used to calculate the resulting evaluations. For evaluation, the FuzzME supports the well-known Mamdani inference [9], Sugeno-WA [11], and Sugeno-WOWA [6] inference. For better description of the latter two methods, see [8].

From the supported inference methods, the Sugeno-WA will be shown as an example. Its advantage is its simplicity (the result is obtained as a weighted average of fuzzy numbers). Another benefit is that the method guarantees that its result is again a fuzzy number (which generally does not hold for the Mamdani inference).

The result of the Sugeno-WA inference [8] is obtained by the following two steps.

1. First, the degree h_i of correspondence between the given m-tuple of fuzzy values $(U'_1, U'_2, \dots, U'_m)$ of partial evaluations and the mathematical meaning of the left-hand side of the i-th rule is calculated for any $i = 1, \dots, n$ in the following way (the minimum operator is used to model the intersection of fuzzy sets)

$$h_i = \min\{hgt(U'_1 \cap U_{i,1}), \dots, hgt(U'_m \cap U_{i,m})\}. \tag{14}$$

2. The resulting fuzzy evaluation U is then computed as a weighted average of the fuzzy evaluations U_i, $i = 1, \dots, n$, which model mathematical meanings of linguistic evaluations on the right-hand sides of the rules, with the weights h_i. This is done by the formula

$$U = \frac{\sum_{i=1}^{n} h_i \cdot U_i}{\sum_{i=1}^{n} h_i}, \tag{15}$$

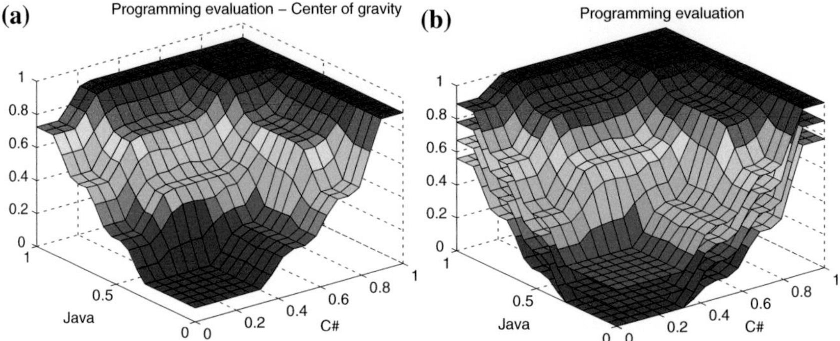

Fig. 9 Evaluation by fuzzy expert system using the SugenoWA inference. **a** Center of gravity. **b** Full visualization of the result

where the multiplication and addition are performed by the standard fuzzy numbers arithmetic operations (i.e. by the Formulas 3 and 4).

The company can also use a fuzzy expert system and describe the relationship between the two aggregated partial evaluations and the resulting evaluation by a set of if-then rules. An example of such a rule might be: *If Java knowledge = good and C# knowledge = very good, then programming knowledge = very good.* Then the resulting evaluation can be calculated using the well-known Mamdani inference, or by the Sugeno-WA inference algorithm.

The Fig. 9b shows the result of the SugenoWA with a sample fuzzy rule base for this example. The information, which fuzzy rules have been used, is not relevant for this chapter. Instead, the figure should show the reader how this evaluating function typically looks like in general. For comparison, only the center of gravity is plotted in the Fig. 9a.

3.10 Interpretation of the Results

The process of aggregation stops when we reach the root of the goals tree. Its evaluation is also the overall evaluation of the given alternative. This evaluation is expressed by a fuzzy number on [0, 1].

The decision-maker is provided with several forms of the final evaluation. All evaluations are presented graphically by means of the graph of the membership function of the evaluation (which is a fuzzy number). Various numerical characteristics are calculated for the evaluations: the center of gravity [4], or the uncertainty measure [8]. Finally, the evaluations are also described verbally by a linguistic approximation on a linguistic variable defined by the decision-maker. This makes the results easy to interpret.

The alternatives can be ordered by the FuzzME software. The centers of gravity of their evaluations are used for this comparison.

The reader is invited to try this simple application in the FuzzME on his/her own. On the http://www.FuzzME.net, the demo-version of FuzzME can be downloaded and the file with this example can be found.

4 FuzzME Software

The software makes it possible to design the fuzzy models of multiple-criteria evaluation based on the described methods. As soon as any change is done in the model, the results are recalculated automatically, so the decision-maker can study the impact of the changes in the model easily. The FuzzME is ideal in situations when a large number of alternatives (even a few thousands) have to be evaluated and when many criteria have to be taken into account. The Fig. 10 shows the main window of the software. On the left-hand side, there is a goals tree editor. The right-hand side differs according to the selected goals tree node. In this case, the figure shows the user interface for setting the normalized fuzzy weights.

The linguistic variables can be designed in the linguistic variable editor (Fig. 11). The user can design the desired linguistic variable easily—when the dialog is opened the number of values and the type of fuzzy numbers (triangular, trapezoidal, etc.) that should model the terms' meanings is chosen. Then the linguistic variable is generated automatically. The user only needs to set the names of the terms. The fuzzy numbers that model the terms' meaning can be of course adjusted by the user.

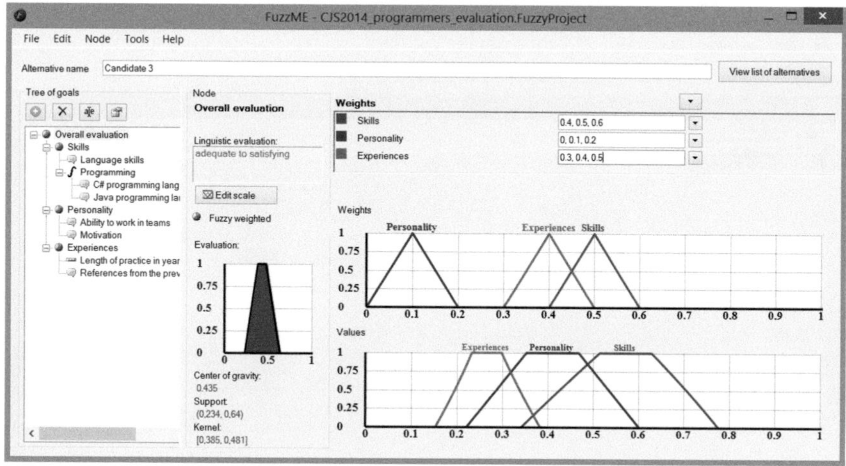

Fig. 10 The main window of the FuzzME software

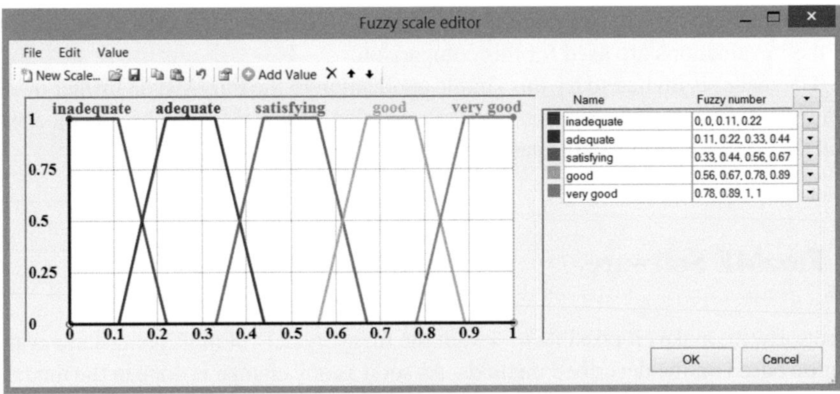

Fig. 11 Designing the linguistic variable in the FuzzME

The results of the evaluation are presented in a comprehensible way (Fig. 12). Next to the alternative name, its fuzzy evaluation (fuzzy number) is plotted. The expert has thus the full information for his/her decision. Below the alternative name, its linguistic description is present. The alternatives can be ordered by their names or by the centers of gravity of their evaluations.

Besides the summary of the results, the expert can view the resulting evaluation of any alternative (which is a fuzzy number on [0, 1]) in the fuzzy number editor. There, the overall evaluation and its characteristics (such as center of gravity, uncertainty measure, etc.) can be studied in detail to make a qualified decision (Fig. 13).

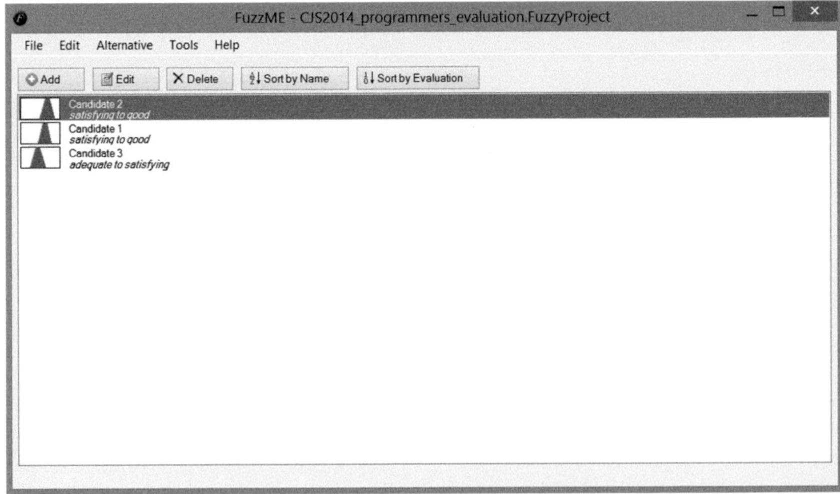

Fig. 12 The summary of the evaluated alternatives in the FuzzME

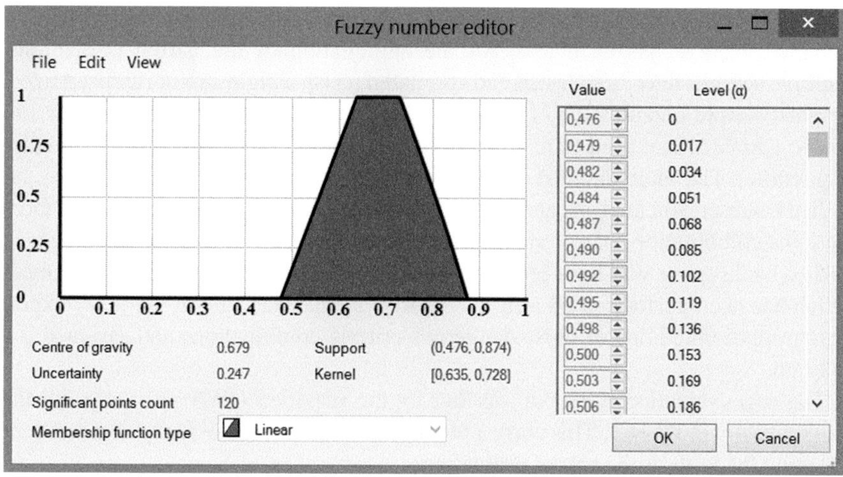

Fig. 13 The alternative's overall evaluation can be viewed in the fuzzy number editor in FuzzME for the more detailed insight into the alternative evaluation

Additional information and the demo-version of the software can be found at the FuzzME website http://www.FuzzME.net.

5 Real-World Applications

Three real-world applications of the FuzzME software will be described. The first one is a soft-fact rating problem of one of the Austrian banks [5]. The results can be then used by the bank in deciding whether a company should be granted a credit or not. The second application concerns with a real programmer evaluation in an IT company [16]. And, finally, the third application presents an assessment of safety for agri-food buildings [1].

5.1 Companies Rating by a Bank

The first application of the FuzzME software was a soft-fact-rating problem of one of the Austrian banks. The problem itself, the way how it has been solved with FuzzME and the conclusions that has been made are described in [5]. This problem was solved in co-operation with the Technical University in Vienna.

The model designed in the FuzzME represented the evaluation of companies according to the soft (qualitative) data. This evaluation, together with evaluation of the companies according to the hard (numerical) data, could be used as one of the materials for deciding on grating a credit by the bank to the company.

The goals tree contained altogether 27 qualitative criteria. First, only the fuzzy weighted average has been used for the aggregation of the partial evaluations. Multiple settings have been tested and compared (crisp weights, or normalized fuzzy weights; simple fuzzification of the scales used by the bank, or more complex linguistic variables that tried to model the meaning of the individual terms as precise as possible). The more detailed description can be found in [5, 8].

In the subsequent analysis and discussion of the results, it has shown out that there are some combinations of criteria that indicate a significant danger that the company will go bankrupt or will have problems acquitting its debt. Therefore, another evaluation has been performed by a fuzzy expert system. Each rule in the fuzzy expert system represented one of those dangerous criteria combinations and assigned it a risk rate.

The two evaluation were put together by the fuzzified OWA operator with the weights $W_1 = \tilde{0}, W_2 = \tilde{1}$. This corresponds to the infimum of the fuzzy numbers representing the both of the partial evaluations.

Altogether, 62 companies were evaluated. This problem clearly showed the advantages of the methodology on which the FuzzME is built—it is possible to use multiple aggregation methods in the same model, the changes in the model can be done easily and, finally, the evaluation of a large number of alternatives is possible. In fact, the subsequent load testing showed that the FuzzME is prepared for much larger set of alternatives than the provided sample. The FuzzME was able to evaluate thousands of companies (generated randomly for purpose of the load test) in a reasonable time.

5.2 Employees Evaluation in an IT Company

The FuzzME was used in the area of HR management for periodic evaluation of employees in the IT company AXIOM SW Ltd [16]. The company has taken a methodology of Microsoft as a base and adjusted it to fit their needs.

The evaluation is based on the so-called competency model, which reflects the competency composition necessary for each of the working roles. The following working roles were identified in the company: senior executive, head of the project, analyst, consultant, software engineer, dealer, and marketing agent. Each of the working roles have different (fuzzy) weights assigned to the particular competencies (such as creative thinking, stress resistance, etc.).

The competencies of individual employees are evaluated by several evaluators—by the employees themselves in the first place, by their direct supervisor, their subordinates (only in case of mangers), and their colleagues working on the same project. The linguistic variables are used, so the evaluation is verbal. The evaluations of the individual evaluators (the fuzzy numbers modeling the meanings of the linguistic variable values) are aggregated by a fuzzy weighted average. Then, the aggregated evaluations of the competencies are again aggregated by a fuzzy weighted average.

The evaluation of specific groups of competencies can be also used to determine the type of the employee. For each of the types, a motivation strategy exists. So the results from the FuzzME need not to be used only for a direct assessment of the employee, but they can be also used to choose the proper motivation strategy for the particular employee.

5.3 *Assessment of Safety in Agri-Food Buildings*

The third example is an assessment of the safety in agri-food buildings described in the paper [1]. The authors of that paper used FuzzME to assess two main aspects of agri-food buildings—the hygienic safety and the workers' safety. Various areas of the buildings are evaluated according to multiple qualitative and quantitative criteria. For aggregation of the evaluations, a fuzzy expert system is used for some of the partial goals and a fuzzy weighted average for the others. The model has been tested on the manufacturing area of a dairy farm located in Italy.

6 Conclusion

In this chapter, a model of multiple-criteria evaluation that is used in the FuzzME software has been described. The theory has been illustrated on a sample application of evaluation of new programmers in a company. The main aim was to show the possibilities of the described methods and the software that implements them. The readers have the possibility to download the described example from the FuzzME website and try it on their own. Finally, three real-world applications of the software have been described.

References

1. Barreca, F., Cardinali, G., Fichera, C.R., Lamberto, L., Modica, G.: A fuzzy-based model to implement the global safety buildings index assessment for agri-food buildings. J. Agric. Eng. **45**(1), 24–31 (2014). doi:10.4081/jae.2014.227
2. Bebčáková, I., Talašová, J., Pavlačka, O.: Fuzzification of Choquet integral and its application in multiple criteria decision making. Neural Netw. World **20**, 125–137 (2010)
3. Choquet, G.: Theory of capacities. Annales de l'institut Fourier **5**, 131–295 (1953)
4. Dubois, D., Prade, H.: Fundamentals of Fuzzy Sets. The Handbook of Fuzzy Sets Series. Kluwer Academic Publishers, Boston (2000)
5. Fürst, K.: Applying fuzzy models in rating systems. Neural Netw. World **20**, 113–124 (2010)
6. Holeček, P., Talašová, J.: FuzzME: a new software for multiple-criteria fuzzy evaluation. Acta Universitatis Matthiae Belii, series Mathematics **16**, 35–51 (2010)
7. Holeček, P., Talašová, J.: Multiple-criteria fuzzy evaluation in FuzzME—transitions between different aggregation operators. In: Talašová, J., Stoklasa, J., Talášek, T. (eds.) Proceedings of the 32nd International Conference on Mathematical Methods in Economics MME 2014, pp. 305–310 (2014)

8. Holeček, P., Talašová, J., Müller, I.: Fuzzy methods of multiple-criteria evaluation and their software implementation. In: Mago, V., Bhatia, N. (eds.) Cross-Disciplinary Applications of Artificial Intelligence and Pattern Recognition: Advancing Technologies, pp. 388–411. IGI Global (2012), ISBN: 978-1-61350-429-1, doi:10.4018/978-1-61350-429-1. ch021
9. Mamdani, E.H., Assilian, S.: An experiment in linguistic synthesis with a fuzzy logic controller. Int. J. Man Mach. Stud. **7**, 1–13 (1975)
10. Pavlačka, O., Talašová, J.: The fuzzy weighted average operation in decision making models. In: Proceedings of the 24th International Conference Mathematical Methods in Economics, pp. 419–426 13–15 Sep 2006. Plzeň (Ed. L. Lukáš)
11. Talašová, J.: Fuzzy methods of multiple criteria evaluation and decision making (in Czech). Publishing House of Palacký University, Olomouc (2003)
12. Talašová, J., Bebčáková, I.: Fuzzification of aggregation operators based on Choquet integral. Aplimat J. Appl. Math. **1**(1), 463–474 (2008). ISSN 1337-6365
13. Torra, V.: The weighted OWA operator. Int. J. Intell. Syst. **12**(2), 153–166 (1997)
14. Yager, R.: On ordered weighted averaging aggregation operators in multicriteria decision making. IEEE Trans. Syst. Man Cybern. **18**(1), 183–190 (1988)
15. Zadeh, L.A.: The concept of linguistic variable and its application to approximate reasoning. Inf. Sci. **8**, 199–249 (1975)
16. Zemková, B., Talašová, J.: Fuzzy sets in HR management. Acta Polytech. Hung. **8**(3), 113–124 (2011)

Part III
Fuzzy Logic in Business and Industrial Practice

Strategic R&D Project Analysis: Keeping It Simple and Smart

Mikael Collan and Pasi Luukka

Abstract Strategic R&D projects require forward-looking analysis and face structural uncertainty, which means that most often precise and detailed information about them is unavailable. This means that any systems that are used in managing them must be robust enough to handle the available imprecise information, while at the same time being simple enough to convey a good-enough overall understanding of these projects. Fuzzy numbers are a precise way of representing imprecise information. Triangular fuzzy numbers are simple to use and have an intuitively understandable graphical presentation. Scorecards are a well-known simple structured tool for the collection and analysis of information. This chapter proposes using triangular fuzzy numbers with scorecards to create a simple, easy to understand, easy to visualize, low-cost, multi-expert analysis tools for strategic R&D projects that can be created by anyone with a laptop computer and spread-sheet software. New weighted averaging operators that are able to handle interdependence between criteria are presented. A numerical example is used to illustrate how a system based on the above-mentioned components works, and how it may offer smart decision-support for the management of strategic R&D projects, under structural uncertainty.

Keywords Strategic R&D · Structural uncertainty · Scorecard · Triangular fuzzy numbers · Imprecision · Aggregation operator

Using fuzzy scorecards to collect data for strategic R&D projects & analyzing and selecting projects with a system that uses new fuzzy weighted averaging operators.

M. Collan (✉) · P. Luukka
School of Business, Lappeenranta University of Technology, Lappeenranta, Finland
e-mail: mikael.collan@Lut.fi

M. Collan et al. (eds.), *Fuzzy Technology*, Studies in Fuzziness and Soft Computing 335, DOI 10.1007/978-3-319-26986-3_10

1 Introduction and Background

Making decisions with regards to choosing, continuing, and shutting down strategic research and development (R&D) projects is a difficult task and sometimes as much of an art as it is a science. Strategic R&D projects may generate cash-flow only sometime in the future, but do not necessarily hold any operational or intrinsic value at the moment. Information available for the analysis of strategic R&D projects is most often imprecise [18]. Imprecision is also present when "near future" R&D projects are considered, but it is of a less dramatic nature. In both these cases having imprecise information makes decision-making more difficult. Yet, still we must make decisions about these projects!

When classifying the type of uncertainty with regards to R&D projects, we can most often talk about two relevant categories, depending on how deep the knowledge (or the lack thereof) is [30, 32]: we can speak about *parametric uncertainty or about structural uncertainty.* We face *parametric uncertainty*, when we understand the possible future states of the world and know the consequences of these states. Parametric uncertainty is what typically faces the decision-maker when decisions are made about R&D projects that will soon be or already could be in operational use. We face *structural uncertainty,* when the "structure of the future" is uncertain, making structural uncertainty a deeper type of uncertainty (than parametric uncertainty). The problem with structural uncertainty is that traditional cash-flow based analysis and single number analysis methods often fail. Structural uncertainty is what most often faces strategic R&D projects and strategic intellectual property rights (IPR) [12].

Potential is something that is usually connected to uncertainty and inaccuracy in estimation that it causes by an inverse relationship—where there is a lot of uncertainty, there is also a possibility that a more positive than the most likely or the "expected" outcome will take place. It is important from the point of view of a strategic R&D decision maker to highlight potential and this is not achieved if estimation inaccuracy is not considered. Single numbers that is, the "normal" numbers that we use every day, do not consider inaccuracy sufficiently. They are "precise" in their representation of information, and do not carry knowledge about the perceived inaccuracy that may, or may not surround them. It is likely that most strategic R&D projects are evaluated by using single numbers, even if simulation and other advanced methods are becoming more popular; this means that important information about the estimation inaccuracy and the negative and the positive potential of strategic R&D projects may be lost. We don't want that to happen, because we lose important information that "we already had" in the first place.

One way to consider and to carry imprecision in analysis is to use fuzzy logic to give a precise expression to the estimation imprecision. Using fuzzy logic [45–47] and fuzzy numbers, a subset of fuzzy sets, is widespread in many industries, and used in many applications ranging from embedded systems within vacuum cleaners to computerized control systems in high speed monorails. For a good overview on applications of fuzzy sets see the Springer book series "Studies on Fuzziness and

Soft Computing" with more than 150 volumes on the subject. Fuzzy sets and fuzzy numbers can also be used in analysis and decision support applications. The "inventor" of fuzzy logic, Lotfi Zadeh thought that decision support and financial applications in general would be among the first of its application areas, it is surprisingly only recently that these areas have widely started to adopt and embrace fuzzy logic. For the purposes of this research we use triangular fuzzy numbers, defined below:

Definition 1 A triangular fuzzy number $\hat{a}$ can be defined by a triplet $\hat{a} = (a_1, a_2, a_3)$. The membership function $\mu_{\hat{a}}(x)$ is defined as [22]

$$\mu_{\hat{a}}(x) = \begin{cases} 0, & x < a_1 \\ \dfrac{x - a_1}{a_2 - a_1}, & a_1 \leq x \leq a_2 \\ \dfrac{x - a_3}{a_2 - a_3}, & a_2 \leq x \leq a_3 \\ 0, & x > a_3 \end{cases} \quad (1)$$

For arithmetic operations for triangular fuzzy numbers we refer to [22].

Fuzzy numbers are the counterpart of fuzzy logic to the "crisp" single numbers that we normally like to use to represent, for example, scorecard scores. Fuzzy numbers are "possibility distributions" that represent information and that illustrate imprecision by the width of the distribution—more imprecision, more width. Our "normal" numbers can be considered special cases of fuzzy numbers—cases with no uncertainty. Triangular fuzzy numbers that are suggested to be used here, due to their simplicity, are a subset of fuzzy numbers and of fuzzy sets in general, and they can be defined with three values (three normal numbers); a_1, a_2, and a_3, where "a_2" is the peak (or center) of the fuzzy number and where a_1 and a_2 represent the minimum possible and the maximum possible values. We use the notation (a_1, a_2, a_3) for this kind of numbers. When the distance from a_1 and a_3 to $a_2 > 0$, the membership function (the function that determines the shape) of the triangular fuzzy number has the form defined in Definition 1 and shown graphically in Fig. 1.

It seems to be a good idea to use fuzzy numbers to not lose the information about estimation uncertainty in our analysis. The next thing is to come up with a simple

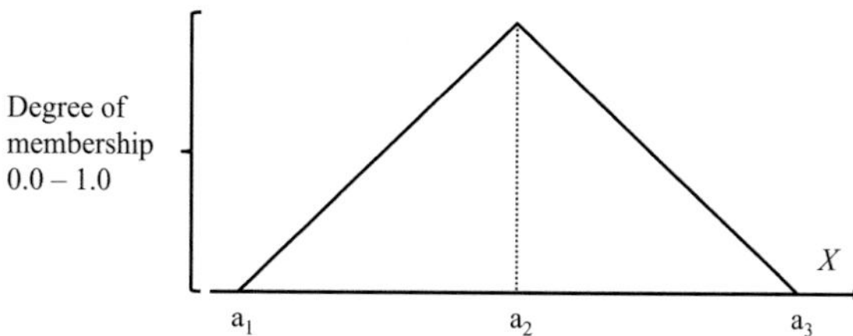

Fig. 1 Graphical presentation of a triangular fuzzy number. X-axis represents the score that, e.g., a criterion receives

and easy to use method to help in the analysis of the strategic R&D projects. Why simple? Well, the main reason for why we don't need a very detail oriented mathematically precise representation is that we don't have the data to support such a model, when we deal with strategic R&D projects that face "close to" structural uncertainty about the future: the detailed information to support and justify the use of such methods just does not exist! It is also most often a good idea not to try to invent (often very complex mathematical) methods that create detailed information for one's purposes (one wonders if that ever happened in academia?). Most often our best bet is to go with more robust information that is available and a simple method that can handle the information as it is, and that at the same time is easy to understand; a common sense requirement for any system that is really put to into production use. For this task we suggest the use of the scorecard. Scorecards have also previously been used in connection with management of R&D, see, e.g. [24, 6] and are likely to be already understood by managers.

From a decision-making science point-of-view the basic form of a scorecard is a simple multiple-criteria decision analysis (MCDA) tool, an elementary weighted, or non-weighted scoring system that uses a numerical scale in scoring a "project" on several given criteria. By scoring parts and adding up the results gives us an idea of the overall situation—"the higher the aggregate score the better the outcome (or the project)". The commonly used scorecards are usually constructed by selecting accurately measurable items as criteria, and they are used by scoring based on measurement results of past performance (facts). A simple scorecard can be used with a pen, while the scorecard is on a piece of paper: just think about writing down your golf round results after each hole on the golf scorecard. The classical scorecard is a backward-looking instrument of performance measurement and it is based on past observed performance. In this environment using crisp numbers, to represent factual history, is not a problem, because the past is most often quite precisely known.

In some cases and especially, when strategic R&D projects are concerned, we are more often than not looking forward and not backward. The information available will be imprecise and no precise measurement is, on many occasions realistically possible, thus the classical scorecard is not a good match, and needs to undergo changes that allow for measurement imprecision in the form of imprecise estimates to be taken into consideration [3]. One can think about how one would need to change a golf scorecard in the case that one would want to beforehand estimate the hole-by-hole score in a way that it would reflect the estimation inaccuracy involved, versus just recording the observed results, and while still keeping things very simple. In this vein a non-trivial observation is that it is very likely that it is the simplicity of the scorecard that has made it such a widespread tool in management: any enhancements should strive to preserve simplicity! One would probably want to include also information in addition to the most expected end result for each hole (number of strokes), such as the best possible score one could play and the worst case score, to show the imprecision of the estimation.

In golf there are many issues that affect the result for each hole, for example, wind, humidity, temperature, condition of the greens, number of hazards, distance to the flag, if one slept well the night before, one's skill in golf, and so on. It is a good bet that we can find dozens of factors that can contribute to the end-result. Now, if we put all of these in a logical order and systematically study their effect and cause relationships, perhaps their joint effect, and make a model and feed estimates of these factors in… then the model would give us an estimate of the end-result for each hole and of subsequently for the whole round. If we start in 15 min we may have a "fresh set of data" to work with—we know how we will feel in 15 min, because we will most likely feel like we feel now, and the weather will most likely remain similar, so we are able to come up with estimates for each factor and we may be satisfied with our estimation accuracy. If the model dependencies and joint effects are correctly set, then we may get a good result from the model that has a low error that is, the difference between the estimated final score for the round and the actual score is small. Still no-one in their right mind expects the model to perform without error. In case the round starts in 3 months from now, estimating the conditions on the course and of the player is more difficult, but perhaps the model can still give us some indication of what to expect.

Now, if the round of golf will take place far away in the future and we don't know what will happen between now and the round, the problem becomes a much more difficult one. Also what becomes under question is the usability of our complex system that is "tuned" for the present day settings. In, for example, 5 years the linkages between factors that now seem logical to us may have changed, or ceased to exist. Perhaps there are new factors that play a role we don't yet know about and maybe… maybe we just don't have a whole lot of precise information overall. It becomes so increasingly difficult to believe in our ability to go into a detailed analysis of the round being played in 5 years from today that we may lose our faith in the model altogether. The reason for this being that the model is too accurate for the job. Does this mean that we cannot systematically estimate the end result of the round? No, it does not, but it means that we need a different tool for the job—a robust one, one that allows us to feel comfortable about the fact that we do not know very accurately. "Not knowing very accurately" does not mean "not knowing at all", we do know a whole lot—but with great inaccuracy present many "small" issues become unimportant and it is the broad picture that we must concentrate on. We must see the forest from the trees! This means that we can use a simple tool and concentrate on estimating only the most important factors to the end result—for example, just the number of strokes played for each hole and as discussed above, perhaps by simply estimating the most expected, the maximum possible and the minimum possible outcomes (scores). Actually, in this way we are even able to roughly estimate our round of golf result far away in the future, on courses we have never played on before… and just based on the course plan that may actually be the only good information we have available at the moment. Interestingly, replacing "round of golf" with "strategic R&D project", "hole" with "criterion affecting the project" tells us a story about estimation of a strategic R&D project.

It is quite clear that the type of scorecard presented, with three scenarios for the future score is highly compatible with using triangular fuzzy numbers. Mapping three different scenarios (a minimum possible-, a most expected-, and a maximum possible outcome) for an input variable with a_1, a_2, and a_3 allows us to create fuzzy number inputs for our strategic R&D project scorecard that we incidentally propose here as a simple tool for the data collection for strategic R&D projects.

Now this may sound as if it is nothing new under the sun, and indeed many have written about fuzzy logic, about scorecards, and about the selection and evaluation of R&D projects. Yet there are not so many out there who talk about fuzzy scorecards for strategic R&D evaluation, and to the best of our knowledge no intelligent systems have been reported built, based on the use of fuzzy scorecards for strategic R&D evaluation. There are many advanced methods, models, and systems for handling R&D project selection and optimizing R&D portfolios that use complex and ingenuous mathematics, see, e.g., [13] for references. The problem with many of the most advanced models, methods, and systems is that unfortunately they are not a good fit with, or at all usable for strategic R&D projects, because they are not robust enough to handle structural uncertainty.

One of the contributions of this chapter, in addition to the aforementioned cross-disciplinary combination of methods, is to show how we do not need a very complicated system with high complexity and incredible mathematics to be able to come up with smart, usable, and intuitively understandable analysis that helps us choose new and manage our existing strategic R&D projects.

Yet another contribution of this chapter is to present new weighted averaging operators: the Fuzzy Heavy Weighted Averaging (FHWA), the Fuzzy Super Heavy Weighted Averaging (FSHWA), and the Fuzzy Super Heavy Ordered Weighted Averaging (FSHOWA) operators. FHWA is an operator that is able to take into account the interdependence, or overlapping, of criteria in the aggregation of (context dependent) fuzzy numbers, and builds on previously introduced Fuzzy Weighted Average (FWA) operator, see e.g. [4] and on Fuzzy Scorecard (FSC) operator [10], treating them as special cases. FSHWA is an operator designed for capturing the type of interdependency that is commonly referred to as synergy between two variables, where the sum of two variables is larger than their single weights alone. FSHOWA is an operator that is designed to extend the well known fuzzy OWA into situations that include synergy. We also present the crisp version Super Heavy OWA (SHOWA).

This chapter continues in the next section, by presenting a fuzzy scorecard for strategic R&D project data collection, and showing how such a scorecard is able to accommodate also changed information. Section three presents the new FHWA, FSHWA, and FSHOWA weighted averaging operators. In section four a system, based on fuzzy scorecards, FHWA, and FSHWA, for the analysis of strategic R&D projects is presented, and illustrated with a numerical example. Finally the chapter closes with a discussion and some conclusions.

2 Fuzzy Scorecard for Strategic R&D Project Data Collection

Above we have described the simple elements of the tool that we are about to introduce, the scorecard and triangular fuzzy numbers. Together these offer a synergy of simplicity and capacity to express imprecision that can, as we show later on, be leveraged to create a rather advanced system for the analysis of strategic R&D projects. Still the scorecard itself is, more than anything else, a platform for collecting information—what comes on top of that is analytical elements that use the information from the scorecard as a starting point. So there is actually a long way from the scorecard to a system that actually optimizes a portfolio of strategic R&D projects.

We start with shortly comparing the classical single number scorecard with the proposed fuzzy "three scenario" scorecard, both visible in Fig. 2. The scorecards shown in Fig. 2 are simplifications of "real world" scorecards, but the simplification does not affect any of their essential characteristics. We can see that the scores given in the classical scorecard are the same as the most expected scores in the fuzzy scorecard—this is not a trivial observation as it means that both include at least the information that is contained in the classical scorecard. Among the benefits of the simple fuzzy scorecard approach is the possibility to offer decision-makers with intuitively understandable graphical presentation of the inputted scores and the imprecision as it is perceived. As a matter of fact, using the triangular fuzzy numbers in scoring allows us to see the connected estimation uncertainty with one

Evaluation Criteria	Classical single number scorecard
Criterion A	3
Criterion B	7
Criterion C	4
Criterion D	8
Criterion E	9
Aggregate score	**31**

Evaluation Criteria	"Three scenario"-fuzzy scorecard scores and graphical presentation of scores		
Criterion A	maximum possible	4	
	most expected	3	
	minimum possible	1	
Criterion B	maximum possible	8	
	most expected	7	
	minimum possible	7	
Criterion C	maximum possible	5	
	most expected	4	
	minimum possible	1	
Criterion D	maximum possible	9	
	most expected	8	
	minimum possible	6	
Criterion E	maximum possible	10	
	most expected	9	
	minimum possible	9	
Aggregate score	**maximum possible**	**36**	
	most expected	**31**	
	minimum possible	**24**	

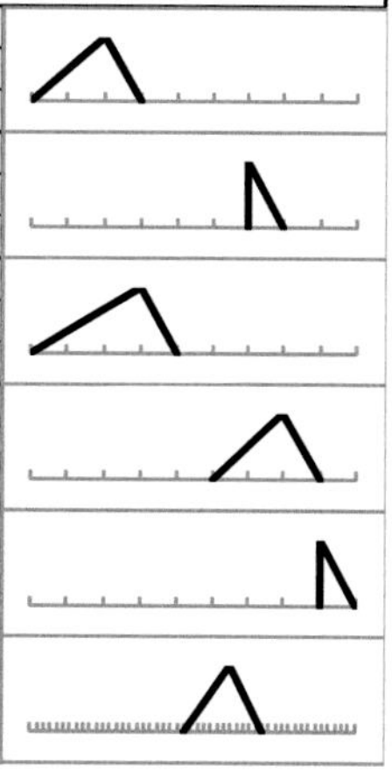

Fig. 2 A single number "classical" scorecard and a "three scenario" fuzzy scorecard

glance. We can see that on many occasions the estimations for the different criteria are asymmetrically distributed around the most expected score. Again this is not a trivial observation, because in the real world a perfectly symmetrical estimation is a rare beast, even if for some reason it is what one "always" finds from the text books. This piece of misinformation about the existence of symmetrical expectations and estimations (and the near non-existence of asymmetric ones) comes from having used such simplifications of reality as "plus-minus ten percent" that are often the basis of methods such as sensitivity analysis, or being a tad too comfortable with the symmetric bell curve of the Gaussian (or normal) distribution, perhaps resulting from using stochastic processes. The bottom line is that the fuzzy scorecard can present symmetric or asymmetric distributions of the scores on equal terms and very simply. This information allows us to see, which criteria have been estimated as having positive potential, and which ones have a downside. The width of the aggregate of the fuzzy estimates shows the overall imprecision in the estimation. The aggregation of the fuzzy scorecard is discussed more in detail below in section three.

It is clear that the "true" relationship of the extreme scenarios (max. and min. possible) with the most expected scenario is most likely different from the linear relationship of the triangular fuzzy number, however for the purposes of most analyses it not so different or "wrong" that it could not be used, and most importantly the reliability of the results will remain at a "good enough" acceptable level. Remembering that under structural uncertainty we may not be able to realistically add a lot of detail into the inputs, as we may not have enough detailed information to support it, makes it easier to accept the use of a simple three scenario based scorecard approach.

Markets for products and services and the development of technology are in a state of constant change, which means that also the evaluations of strategic R&D projects should change with the markets. Projects that are important may become less important, new information may be revealed and imprecision becomes smaller, old information may become compromised and imprecision may increase. What this means in practice is that strategic R&D projects should be and in practice are evaluated not once, but many times during their life.

As they are revisited the changes in the information should be visible in the estimates that is, the estimates should follow the evolution of the information. In Fig. 3 we can see the evolution of the score of "Criterion A", where the best estimate scenario remains unchanged, but as information quality increases the imprecision is gradually lifted and when no imprecision remains (at time 2) the estimate reduces to a single number estimate.

This shows how a simple triangular estimate can convey information about the direction of the evolution of a score. It is important to note that if one uses a single number estimate, none of this information is visible, as the most expected score remains unchanged (the single number remains unchanged). The direction into which different criteria, on which a strategic R&D project is estimated, are evolving may be very important information from the point of view of decision-making. Especially a change in direction/trend may be important. Power of graphical

time 0	maximum possible	4	
	most expected	3	
	minimum possible	1	
time 1	maximum possible	3	
	most expected	3	
	minimum possible	2	
time 2	maximum possible	3	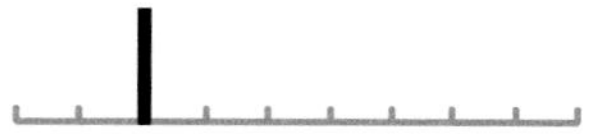
	most expected	3	
	minimum possible	3	

Fig. 3 Evolution of the score of criterion A through time from t = 0 to t = 2

presentation in understanding the change intuitively cannot be emphasized too much. The most expected score can naturally also change, but above it did not, to illustrate the point made.

By using the above explained simplistic approach, we can very easily build a structured way of collecting estimates of strategic R&D projects and for keeping track of the evolution of the estimates, on a criterion-by-criterion-level and naturally also on the aggregate score-level. Collecting data and looking at the way it is evolving is one thing, using the data further to create advanced analysis for managing strategic R&D projects is another. Next we turn to presenting new weighted averaging operators, FHWA, FSHWA, SHOWA, and FSHOWA that are able to handle scorecards with weighted criteria for considering overlap between information that cause them to have a joint weight that is lower than their weight separately, and synergy between criteria that cause their joint weight to be higher than the sum of their separate weights.

3 New Weighted Averaging Operators for Handling Redundant and Interdependent Information

Research that is based on aggregating information based on the notion of weighted averaging (WA) has blossomed with an impressive number of academic research published on a number of variants of weighted averaging operators, and on a myriad of applications of information aggregation. For introduction to aggregation functions and operators for aggregating information we refer suggest the interested reader see [2, 39]. As already discussed above, we are mostly interested in weighted averaging of values represented as fuzzy numbers. Fuzzy weighted averaging operator can be found from many sources in the literature, but for a definition we refer here to [4]. Earlier work on applying fuzzy weighted averaging includes, for example, [17, 21, 33, 35]. Also alternative approaches for the fuzzy weighted average have been presented, see [36].

An active area of research within the field of weighted averaging operators are Ordered Weighted Averaging (OWA) operators, introduced by [40]. Several different types of versions of the original OWA have been introduced, and include, for example, generalized OWA (GOWA) [31, 42], Ordered Weighted Geometric Averaging operator (OWGA) [26, 39], Induced Ordered Weighted Averaging Operator (IOWA) [9, 41], linguistic OWA operators [19, 38], Fuzzy OWA (FOWA) [7, 27], and Heavy OWA [44]. There are also many others who have contributed to the literature on OWA operators with a variety of extensions.

Besides the different versions of OWA, emphasis has also been given to the determination of the weights used in connection with OWA operators. Here some of the more popular approaches include quantifier guided aggregation presented in [40] and a procedure to generate OWA weights that have a predefined degree of orness and that maximize the entropy (of the OWA weights) presented in [28, 29]. An analytical solution for O'Hagan's procedure by using Lagrange multipliers was presented by [15]. Filev and Yager [14] developed two procedures based on exponential smoothing to obtain the OWA weights. Fullér and Majlender [16] introduced a minimum variance method (MVM) that follows a concept similar to the O'Hagan's procedure. Wang, Luo, and Liu proposed a chi-square method (CSM) [37] for generating the OWA weights by minimizing dispersion. Obtaining the weights for OWA can be considered to be a mixed integer linear programming problem, solvable by algorithms as was done, for example, by Carlsson and others [8]. OWA operators have also attracted a wide range of applications, see for example [39, 43].

Of the above mentioned works on weighted averaging operators only one addresses the problem of overlapping/redundant information (or fusion of information): the paper by Yager that presents the Heavy OWA operator [44]. This means that this subject of "interaction between variables" in the context of weighted averaging operators merits more study and this is the angle from which we approach the subject. We assume that the aggregated (averaged) elements are predominantly fuzzy numbers, and secondly we assume that situations, where information contained in variable values can be overlapping or interdependent exist.

3.1 Fuzzy Heavy and Super Heavy Weighted Averaging

We go about introducing the new operators by way of first introducing two "old" known operators for reference, and then we introduce our new Fuzzy Heavy Weighted Averaging (FHWA) operator, that considers the two presented old operators as special cases. We start by introducing the first previously presented operator, the Fuzzy Weighted Averaging (FWA) and give the definition for it in the following:

Definition 2 Let U be the set of fuzzy numbers. A Fuzzy Weighted Averaging (FWA) operator of dimension n is a mapping FWA: $U^n \rightarrow U$ that has an associated weighting vector W of dimension n given as

$$FWA(\hat{a}_1, \hat{a}_2, \ldots, \hat{a}_n) = \frac{\alpha_1 \hat{a}_1 + \alpha_2 \hat{a}_2 + \ldots + \alpha_n \hat{a}_n}{\alpha_1 + \alpha_2 + \ldots + \alpha_n} = \sum_{i=1}^{n} w_i \hat{a}_i \quad (2)$$

where $w_i = \frac{\alpha_i}{\alpha_1 + \alpha_2 + \ldots + \alpha_n}$, $i = 1,\ldots,n$ and $\sum_{i=1}^{n} w_i = 1$.

The FWA operator is usable, when different criteria are used to analyze a single object, and when each criterion is assumed to partially explain the object (score). In such a case the criteria are averaged by using the weights to characterize the relative importance of each criterion.

Another previously introduced operator, relevant here and already discussed above, is the Fuzzy Score Card (FSC) operator that can be defined as:

Definition 3 Let U be the set of fuzzy numbers. A Fuzzy Score Card (FSC) operator of dimension n is a mapping: $U^n \rightarrow U$ such that:

$$FSC(\hat{a}_1, \hat{a}_2, \ldots, \hat{a}_n) = \sum_{i=1}^{n} \hat{a}_i \quad (3)$$

When using the classical scorecard or the fuzzy scorecard, presented above, the aggregation over the criteria is made by assuming the independence of the criteria from each other. In other words, this means that the information that each estimated criteria carries is believed to be distinct and non-redundant. We refer to [10] for more information about the fuzzy score card.

Next we turn to the new fuzzy aggregation operator and present the definition for the Fuzzy Heavy Weighted Averaging (FHWA) operator:

Definition 4 Let U be the set of fuzzy numbers. A Fuzzy Heavy Weighted Averaging (FHWA) operator of dimension n is a mapping FHWA: $U^n \rightarrow U$ that has an associated weighting vector W of dimension n such that the sum of the weights is between $[1,n]$ and $w_i \in [0, 1]$, then:

$$FHWA(\hat{a}_1, \hat{a}_2, \ldots, \hat{a}_n) = \sum_{i=1}^{n} w_i \hat{a}_i \quad (4)$$

where $(\hat{a}_1, \hat{a}_2, \ldots, \hat{a}_n)$, are now fuzzy triangular numbers of form given in Definition 1. This leads to our proposition:

Proposition 1 *The FWA and FSC operators are special cases of the FHWA operator, when we are at the boundaries of the FHWA weighting vector.*

We get the FSC operator in the case, when $\sum_{i=1}^{n} w_i = n$, since we have $w_i = 1 \quad \forall i$, because $w_i \in [0, 1]$. In the case of the other boundary, $\sum_{i=1}^{n} w_i = n$, and the requirement of FWA are met because $w_i \in [0, 1]$.

Definition 5 Let U be the set of fuzzy numbers as in Definition 1. A Fuzzy Super Heavy Weighted Averaging (FSHWA) operator of dimension n is a mapping FSHWA: $U^n \rightarrow U$ that has an associated weighting vector W of dimension n such that the sum of the weights is between [1, kn] and $w_i \in [0, k]$, then:

$$FSHWA(\hat{a}_1, \hat{a}_2, \ldots, \hat{a}_n) = \sum_{i=1}^{n} w_i \hat{a}_i \tag{5}$$

In the above definition we can see that the size (weight) of the positive synergy has been limited at the size of the added weight of the independent criteria involved that is, the joint weight can be at maximum k times the size of the added weight of the independent criteria involved. The limitation can be relaxed and the positive synergy can be expected to grow infinitely, which leads to the following definition:

Definition 6 Let U be the set of fuzzy numbers as in Definition 1. A Fuzzy Super Heavy Weighted Averaging (FSHWA) operator of dimension n is a mapping FSHWA: $U^n \rightarrow U$ that has an associated weighting vector W of dimension n such that $\sum_{i=1}^{n} w_i > 1$ and $w_i \geq 0$, then:

$$FSHWA(\hat{a}_1, \hat{a}_2, \ldots, \hat{a}_n) = \sum_{i=1}^{n} w_i \hat{a}_i \tag{6}$$

As with the FHWA operator, assigning the weights correctly can play an important role with FSHWA.

We have above seen the types of situations, where the FWA and the FSC operators are useful, now we explain the intuition behind the case, when the sum of weights is between [1,n], and that can be considered with FHWA. Consider, for example, a case presented in [25], where a summer trainee is evaluated, based on five criteria (needs of the department): (1) Statistical data-analysis skills; (2) Internet data collection skills; (3) Written English language reporting capability; (4) Assisting skills in interviews, and (5) Conference organization skills. When we analyze the five criteria, we can quite safely assume that statistical data-analysis skills do not depend on the other criteria and thus gives distinct information about the candidates; we can assign this criterion the weight of one, reflecting the criteria independence. Criteria 3 and 4, on the other hand, both depend on how well the candidate masters the English language, while at the same time having excellent English language skills does not guarantee a high evaluation for these two criteria. In fact, we have a case of partially overlapping information, where we cannot assume that the two criteria are fully distinct and non-redundant, while we cannot use an averaging operator to combine the two criteria into one criterion, because the criteria are partially distinct. It is in this kind of real world context that the FHWA shows its usefulness, since by setting the sum of the weights between [1,n] and requiring that $w_i \in [0, 1]$ we can consider partially overlapping information.

In the case of strategic R&D projects there may be several similarly overlapping issues that may simultaneously affect many criteria used in the evaluation. Such issues may include availability of expert workforce, former experiences with a certain type of projects, or other firm specific underlying factors. Competencies and other factors with regards to R&D projects may also create synergies, for example, a high level of staff competence together with high quality equipment may render better possibilities for success than either one of these properties separately can contribute to success. Synergies are known to exist in mergers and acquisitions (M&A) and grow from the joint strengths of adjoining companies, who become more powerful together than as separate parts, for references about M&A synergies see, for example, [5, 11, 20].

3.2 New Ordered Weighted Aggregation Operators

Based on the original work of Yager [40] on ordered weighted averaging operators (OWA) and discussed further, e.g., in [34] and based on Yager's later work [44] on heavy ordered weighted averaging operators (HOWA), we introduce two new operators that extend the HOWA operator to consider the situation, where positive synergy among criteria exists. Thus, we define the new Super Heavy Ordered Weighted Averaging (SHOWA) operator, where we also consider the ordering component included in the OWA based aggregation operators:

Definition 8 Let U be the set of crisp numbers. A Super Heavy Ordered Weighted Averaging (SHOWA) operator of dimension n is a mapping SHOWA: $U^n \to U$ that has an associated weighting vector W of dimension n such that $\sum_{i=1}^{n} w_i > 1$ and $w_i \geq 0$, then:

$$SHOWA(a_1, a_2, \ldots, a_n) = \sum_{i=1}^{n} w_i b_i \tag{7}$$

where b_i is the ith largest element in the collection $a_1, a_2, \ldots, a_n$.

The SHOWA operator is monotonic and commutative operator. It is monotonic, because if $a_i \geq e_i \forall i$ then $SHOWA(a_1, \ldots, a_n) \geq SHOWA(e_1, \ldots, e_n)$. It is also commutative, because any permutation of the arguments has the same evaluation. Meaning here that $SHOWA(a_1, \ldots, a_n) = SHOWA(e_1, \ldots, e_n)$, where $(e_1, \ldots, e_n)$ is any permutation of the arguments $(a_1, \ldots, a_n)$. It is also bounded by the open interval $(\min(a_1, \ldots, a_n), \infty)$.

For the SHOWA we can also analyze the magnitude, (a term used and defined by Yager in [44] for the sum elements in the weighting vector W), of the weighting vector $|W| = \sum_{i=1}^{n} w_i$. In order to normalize this feature of the W a characterizing parameter β value of the vector W is introduced. For SHOWA operator this can be defined as $\beta(W) = (|W| - 1)/(n - 1)$. Now, if $|W| = n$, then we have the same magnitude that we would have for the (fuzzy) scorecard operator, but we will not necessarily obtain the same aggregated value. Following [44], and after having analyzed the magnitude of $|W|$, we can further examine the character of the weight vector with four characterization measures. The first characterization measure, the "*attitudinal character*", can be defined as:

$$\alpha(W) = \frac{1}{|W|(n-1)} \sum_{i=1}^{n} (n-i)w_i \tag{8}$$

where we have $\alpha(W) \in [0, 1]$. The second characterization measure, the "*entropy of dispersion*", can be defined as:

$$E(W) = -\frac{1}{|W|} \sum_{i=1}^{n} w_i ln\left(\frac{w_i}{|W|}\right) \tag{9}$$

The third characterization measure, which is often called "*divergence*" of W, would similarly be:

$$Div(W) = \frac{1}{|W|} \sum_{i=1}^{n} w_i\left(\frac{n-i}{n-1} - \alpha(W)\right)^2 \tag{10}$$

The fourth characterization measure, also often used with the OWA-operators, is the "*balance*" operator which is defined as:

$$BAL(W) = \frac{1}{|W|} \sum_{i=1}^{n} w_i\left(\frac{n+1-2i}{n-1}\right) \tag{11}$$

What can be noted is that all these four characterization measures are reduced to the usual definitions for the OWA operator, when $|W| \rightarrow 1$.

Furthermore, Yager [44] defined a measure for *redundancy* ρ in HOWA as follows:

$$\rho = 1 - \beta = \frac{n - |W|}{n-1} \tag{12}$$

For a scorecard type operator, with $W = [1,1]$, we get the redundancy $\rho = 0$ *(no redundancy),* at the other extreme, when we are dealing with a weighted averaging operator, with $W = [\alpha, 1 - \alpha]$, we get the redundancy $\rho = 1$ *(total redundancy)*. For a case of partial redundancy, for example with $W = [1,0.5]$, we get the redundancy $\rho = 0.5$.

The redundancy measure can be also used in connection with the SHOWA operator, but in the event of synergy the redundancy measure may exhibit "unintuitive behavior": in the case of synergy and weight values over one we can have, for example, $W = [1.5,1.5]$, and we get redundancy of $\rho = -1$. The redundancy measure can be also used as a measure of synergy, where negative values of redundancy would indicate the existence of synergy, however, for reaching a more intuitive representation "larger the synergy—larger the value" we introduce a *synergy* measure S that is defined as:

$$S = \frac{|W| - n}{n - 1} \tag{13}$$

For a weighted averaging operator $W = [\alpha, 1 - \alpha]$ we get a synergy value $S = -1$, for a case of partial redundancy, with $W = [1,0.5]$ we get $S = -0.5$, and for a scorecard type aggregation, with no synergy, but with zero redundancy, with $W = [1,1]$ we get $S = 0$. In the previously shown case with synergy, with $W = [1.5,1.5]$ we get $S = 1$.

Next, we define a new super heavy OWA operator for fuzzy numbers:

Definition 9 Let U be the set of fuzzy numbers as in Definition 1. A Fuzzy Super Heavy Ordered Weighted Averaging (FSHOWA) operator of dimension n is a mapping FSHOWA: $U^n \rightarrow U$ that has an associated weighting vector W of dimension n, such that $\sum_{i=1}^{n} w_i > 1$ and $w_i \geq 0$, then:

$$FSHOWA(\hat{a}_1, \hat{a}_2, \ldots, \hat{a}_n) = \sum_{i=1}^{n} w_i \hat{b}_i \tag{14}$$

where $\hat{b}_i$ is the ith largest element in the collection $\hat{a}_1, \hat{a}_2, \ldots, \hat{a}_n$

Proposition 2 *In the case that positive synergy exists among criteria, there has to be at least one individual weight $w_i > 1$ in the super heavy operators.*

Now, since the weighting vector W does not change from the previous definition, we can see that magnitude $|W|$ stays the same as with the SHOWA. This also indicates that the four attitudinal characters do not change. Also the monotonicity, the commutativity, and the boundedness are the same for FSHOWA and for SHOWA. If $|W| = n$ we get the same magnitude as for the fuzzy scorecard (FSC) operator, but not necessarily the same aggregated value.

In the next section, we turn to building an advanced system to handle information gathered with fuzzy scorecards, from multiple experts simultaneously. Gathering information from multiple experts for the same R&D projects will most likely contribute to the reliability and the consistency of the estimates, while at the same time requiring a method to consolidate the estimates into joint estimates. This may also mean finding consensus of estimates, however, consensual dynamics is a topic left outside of the scope of this chapter. We use the FHWA operator in the aggregation of the information over the used criteria. Furthermore, the numerical illustration of the SHOWA and the FSHOWA operators are left as "food for thought" for further research.

4 Simple Multi-Expert System for Strategic R&D Project Ranking that Uses Fuzzy Scorecards, FHWA, and FSHWA

Based on the above discussion about the usability of fuzzy scorecards as a basis for creating intelligent systems for analysis and evaluation of strategic R&D projects, we propose a system that is able to integrate multiple experts' evaluations of a set of projects (using a fuzzy scorecard), and that aggregates their scores with the new FHWA and FSHWA operators, to yield fuzzy scores for the projects. The fuzzy scores are ordered to create a ranking of the projects that can be used, e.g., when portfolio selection is made.

4.1 General Description of the Method

We consider the following general situation, where a finite set of alternatives $A=\{A_i|i=1, \ldots, m\}$ needs to be evaluated by a committee of decision-makers $D=\{D_l|l=1,2, \ldots, k\}$, by considering a finite set of given criteria $C=\left\{C_j|j=1,2, \ldots, n\right\}$. A decision matrix representation of performance rating of each alternative A_i is considered, with respect to each criterion C_j as follows:

$$X=\begin{bmatrix} x_{11} & \cdots & x_{1n} \\ \vdots & \ddots & \vdots \\ x_{m1} & \cdots & x_{mn} \end{bmatrix} \tag{15}$$

where m rows represent m possible candidates, n columns represent n relevant criteria, and x_{ij} represents the performance rating of the ith alternative, with respect to jth criterion C_j. These ratings are obtained by using fuzzy score cards, and are triangular fuzzy numbers. We use the FWA operator to aggregate the fuzzy decision matrixes from each decision maker to one single decision matrix, and simply calculate an average evaluation over the decision makers, with the weighting vector $w_l=\frac{1}{k}, \forall l$. After the creation of the joint triangular ratings for each criterion, the next step is to form a linear scale transformation of the decision matrix, to transform the various criteria scales into comparable scales. If necessary, the criteria set can be divided into benefit criteria (larger the rating, the greater the preference) and into a cost criteria (the smaller the rating, the greater the preference). The resulting normalized fuzzy decision matrix can be represented as:

$$R=\left(r_{ij}\right)_{m\times n} \tag{16}$$

where B and C are the sets of benefit criteria and cost criteria, respectively, and

$$r_{ij} = \left(\frac{a_{ij}}{c_j^{\oplus}}, \frac{b_{ij}}{c_j^{\oplus}}, \frac{c_{ij}}{c_j^{\oplus}} \right) \; j \in B$$

$$r_{ij} = \left(\frac{a_j^{\ominus}}{c_{ij}}, \frac{a_j^{\ominus}}{b_{ij}}, \frac{a_j^{\ominus}}{a_{ij}} \right) \; j \in C$$

where $c_j^{\oplus} = \max_i (c_{ij}), j \in B$ and $a_j^{\ominus} = \min_i (a_{ij}), j \in C$.

This normalized decision matrix is then aggregated with regards to the criteria by using FHWA:

$$R_i = FHWA(r_{i1}, r_{i2}, \ldots, r_{in}) = \sum_{j=1}^{n} w_j r_{ij} \tag{17}$$

$w_j \in [0, 1]$ and $\sum_{j=1}^{n} w_j \in [1, n]$. The selection of weight now needs to be done by considering how much each individual criterion has overlapping information with other criteria, and by this way getting a weight $w_j \in [0, 1]$, where a weight of 1 means fully distinct non-redundant information, and the further the obtained weight is from 1, the more overlap the criterion has with other criteria. The weighting is context dependent, and must be done separately, case by case. The result from the aggregation is a triangular fuzzy number score for each alternative.

The final step in the proposed system is the ordering of the fuzzy scores. There are a number of methods for the ordering of fuzzy numbers, but we refer here to the method introduced by Kaufmann and Gupta [23]. Next, we illustrate the proposed system with a numerical example.

4.2 Numerical Example

Suppose that a company has funding to start one strategic R&D project for a certain field of business; after a preliminary screening six competing projects $A_1, A_{2,\ldots,} A_6$ remain. Five selected benefit criteria are considered for each project:

(1) C_1—*Intellectual capital potentially emanating from the project*
(2) C_2—*Fit with market strategy*
(3) C_3—*Expected competitive advantage*
(4) C_4—*Technological familiarity (to us)*
(5) C_5—*Fit with technology strategy*

Evaluations have been made by three experts. The proposed method is applied to solve the problem and the computational procedure is summarized as follows:

Step 1: Each expert uses a fuzzy scorecard to evaluate the projects with respect to the five criteria. Separate multiple experts can also be used for each criterion, if that is deemed necessary. The ratings for all criteria are presented in Table 1.

Step 2: Calculating the average evaluations of the experts by using FWA. Results given in Table 2.

Step 3: Construction of the normalized fuzzy decision matrix, visible in Table 3.

Step 4: Selection of the weights for each criterion, based on how much overlapping information they contain. Here the weighting vector used is $w = [1, 1, 0.7, 0.6, 0.6]$; we assume that the first two criteria provide fully distinct, non-redundant information and the three remaining criteria are partly overlapping. Here the technology focus of the firm serves the market focus. Then the decision matrix is aggregated using the FHWA over all criteria, with the chosen weighting vector. The resulting aggregated values are visible in Table 4.

Step 5: Finding a linear order of the resulting fuzzy numbers. This is done by using the method introduced by Kaufmann and Gupta [23]; results can be found in Table 5.

Table 1 Project ratings from decision makers; summary for all criteria

DM_1	C_1	C_2	C_3	C_4	C_5
A_1	(4, 7, 9)	(7, 9, 10)	(3, 5, 8)	(9, 10, 10)	(3, 5, 6)
A_2	(7, 9, 10)	(9, 10, 10)	(7, 8, 10)	(9, 10, 10)	(8, 9, 10)
A_3	(9, 10, 10)	(5, 7, 8)	(7, 8, 10)	(5, 8, 10)	(7, 9, 10)
A_4	(4, 7, 9)	(3, 5, 8)	(7, 9, 10)	(9, 10, 10)	(3, 5, 6)
A_5	(4, 7, 9)	(9, 10, 10)	(7, 9, 10)	(9, 10, 10)	(7, 9, 10)
A_6	(3, 5, 8)	(9, 10, 10)	(9, 10, 10)	(4, 7, 9)	(4, 7, 9)
DM_2	C_1	C_2	C_3	C_4	C_5
A_1	(5, 6, 7)	(4, 5, 6)	(5, 6, 8)	(6, 8, 9)	(5, 6, 7)
A_2	(8, 9,10)	(3, 5, 7)	(7, 8, 9)	(8, 9, 10)	(6, 7, 9)
A_3	(5, 6, 7)	(7, 9, 10)	(6, 7, 8)	(9, 10, 10)	(8, 9, 10)
A_4	(5, 6, 8)	(6, 7, 8)	(8, 9, 10)	(6, 8, 9)	(6, 7, 8)
A_5	(6, 7, 8)	(3, 4, 5)	(7, 8, 9)	(7, 8, 9)	(8, 9, 10)
A_6	(5, 7, 8)	(5, 6, 8)	(8, 9, 10)	(7, 8, 9)	(3, 5, 7)
DM_3	C_1	C_2	C_3	C_4	C_5
A_1	(4, 5, 7)	(5, 6, 7)	(4, 6, 7)	(5, 7, 8)	(4, 6, 8)
A_2	(5, 6, 7)	(4, 5, 6)	(6, 7, 9)	(9, 10, 10)	(6, 9, 10)
A_3	(3, 5, 6)	(9, 9, 10)	(5, 6, 9)	(6, 7, 9)	(7, 8, 9)
A_4	(6, 7, 9)	(5, 7, 9)	(7, 8, 9)	(6, 7, 8)	(5, 6, 7)
A_5	(5, 8, 9)	(4, 5, 7)	(6, 7, 8)	(6, 8, 10)	(7, 8, 9)
A_6	(6, 7, 8)	(6, 7, 8)	(7, 9, 10)	(6, 8, 10)	(4, 5, 6)

Table 2 Decision matrix

	C_1	C_2	C_3	C_4	C_5
A_1	(4.3, 6, 7.7)	(5.3, 6.7, 7.7)	(4, 5.7, 8)	(6.7, 8.3, 9)	(4, 5.7, 7)
A_2	(6.7, 8, 9)	(5.3, 6.7, 7.7)	(6.7, 7.6, 9.3)	(8.7, 9.7, 10)	(6.7, 8.3, 9.7)
A_3	(5.7, 7, 7.7)	(7, 8.3, 9.3)	(6, 7, 9)	(6.7, 8.3, 9.7)	(7.3, 8.7, 9.7)
A_4	(4.7, 6, 7.7)	(5, 7, 8.7)	(6, 7.3, 9)	(6.3, 8, 9)	(6.7, 7.7, 8.3)
A_5	(6, 8, 9)	(3.7, 5.3, 7)	(7.3, 8.3, 9)	(6.7, 8.3, 9.7)	(8, 9, 9.7)
A_6	(5, 7, 8.3)	(4.7, 6, 8)	(8, 9.3, 10)	(7.3, 8.7, 9.7)	(3.7, 5.7, 7.3)

Table 3 The normalized decision matrix

	C_1	C_2	C_3	C_4	C_5
A_1	(0.43, 0.6, 0.77)	(0.53, 0.67, 0.77)	(0.4, 0.57, 0.8)	(0.67, 0.83, 0.9)	(0.4, 0.57, 0.7)
A_2	(0.67, 0.8, 0.9)	(0.53, 0.67, 0.77)	(0.67, 0.77, 0.93)	(0.87, 0.97, 1)	(0.67, 0.83, 0.97)
A_3	(0.57, 0.7, 0.77)	(0.7, 0.83, 0.93)	(0.6, 0.7, 0.9)	(0.67, 0.83, 0.97)	(0.73, 0.87, 0.97)
A_4	(0.47, 0.6, 0.77)	(0.5, 0.7, 0.87)	(0.6, 0.73, 0.9)	(0.63, 0.8, 0.9)	(0.67, 0.77, 0.83)
A_5	(0.6, 0.8, 0.9)	(0.37, 0.53, 0.7)	(0.73, 0.83, 0.9)	(0.67, 0.83, 0.97)	(0.8, 0.9, 0.97)
A_6	(0.5, 0.7, 0.83)	(0.47, 0.6, 0.8)	(0.8, 0.93, 1)	(0.73, 0.87, 0.97)	(0.37, 0.57, 0.73)

Table 4 Aggregated values for the candidates by using FHWA

Project	FHWA(C_1, C_2, C_3, C_4, C_5)
A_1	(1.89, 2.50, 3.05)
A_2	(2.59, 3.08, 3.50)
A_3	(2.53, 3.04, 3.49)
A_4	(2.17, 2.75, 3.30)
A_5	(2.36, 2.96, 3.39)
A_6	(2.19, 2.81, 3.35)

Table 5 Ordering of the resulting fuzzy scores

Project	Removal	Divergence	Mode	Order
A_1	2.49	1.17	2.50	6
A_2	3.06	0.91	3.08	1
A_3	3.03	0.96	3.04	2
A_4	2.74	1.13	2.75	5
A_5	2.92	1.03	2.96	3
A_6	2.79	1.17	2.81	4

We find that based on the experts' evaluations the best project is project A_2.

Next, we consider the synergy effect in our computations; criteria one and two have a clear synergy. This changes the used weighting vector into $w = [2, 2, 0.7, 0.6, 0.6]$. For considering the synergy, we apply the Fuzzy Super Heavy Weighted Average (FSHWA) operator. Now re-doing steps 4 and 5 of the process gives us the results presented in Tables 6 and 7.

Table 6 Aggregated values for the candidates by using FSHWA

Project	FSHWA(C_1, C_2, C_3, C_4, C_5)
A_1	(3.79, 4.58, 5.19)
A_2	(3.79, 4.55, 5.17)
A_3	(3.32, 4.29, 4.99)
A_4	(3.15, 4.11, 4.99)
A_5	(3.13, 4.05, 4.94)
A_6	(2.85, 3.77, 4.59)

Table 7 Ordering of the resulting fuzzy scores

Project	Removal	Divergence	Mode	Order
A_1	3.74	1.73	3.77	6
A_2	4.51	1.38	4.55	2
A_3	4.53	1.40	4.58	1
A_4	4.04	1.80	4.05	5
A_5	4.22	1.66	4.29	3
A_6	4.09	1.83	4.11	4

Applying the FHWA in the aggregation results in the order $A_2 \prec A_3 \prec A_5 \prec A_6 \prec A_4 \prec A_1$, where A_2 is the best project. When we apply the FSHWA, and take the synergy into consideration, the resulting order is $A_3 \prec A_2 \prec A_5 \prec A_6 \prec A_4 \prec A_1$, where A_3 is the best project.

The difference in the result can be attributed to the "synergy" effect. Obviously we want to stress that the proper selection of weights is paramount in obtaining credible and usable results.

5 Discussion and Conclusions

Strategic R&D projects often face structural, or near-structural, uncertainty that causes detailed or precise information about the future to be unavailable. This means that tools that we employ in analyzing these projects must built in a way that they can handle imprecision that is omnipresent in forward-looking estimation, and that commonly comes from experts and managers. At the same time management tools should be robust, simple to use and easy to understand. For these reasons we have suggested the joint use of scorecards, a simple method for a structured collection of estimates and triangular fuzzy numbers, a way to include expert estimation imprecision in the analysis. This combo seems to offer a good fit with the analysis of strategic R&D projects. We illustrated the construct of the classical and the simple fuzzy scorecards.

Aggregating the scorecard information is an important issue, because different ways of aggregating the collected information have different effects on the type of results one gets, and that are later used as decision-support. One way of aggregating information, and we feel a suitable way for information gathered through the use of scorecards, is the use of weighted averaging based operators. For this reason we have presented new weighted averaging operators, the fuzzy heavy weighted averaging (FHWA), super heavy weighted averaging (SHWA), fuzzy super heavy weighted averaging (FSHWA), and crisp and fuzzy versions of super heavy ordered weighted averaging (SHOWA, FSHOWA) operators that are able to consider the interaction between variable values (two pieces of information): overlap of information and information that exhibits synergy effects.

We have proposed that the previously known operators, fuzzy weighted averaging (FWA) and fuzzy scorecard (FSC) operator are special cases of the FHWA. The FHWA operator and the four "super heavy" weighted averaging operators are theoretically new contributions, and we feel that there are many avenues of further research especially within researching the scope of applications of these operators.

Under structural uncertainty we do not often have information that would justify the use of complex systems that require detailed information, hence we need simpler more robust systems. Based on this observation and on the available data, collected with fuzzy scorecards, we have proposed a rather robust system for the multi-expert multiple-criteria ranking of strategic R&D projects that is able to accommodate existing information about overlapping information that underlies the evaluation criteria used by using the FHWA operator. The proposed method was numerically illustrated to show that starting from imprecise information collected with fuzzy scorecards, a rather intelligent ranking system can be easily built. Furthermore, a numerical illustration of using the FSHOWA operator showed that including the synergy effect may have benefits for such problems that truly exhibit synergy. It is important to note that the proper selection of weights is paramount in obtaining credible and usable results, when using any weighted averaging based aggregation methods. Weight selection is an interesting avenue for further research.

As a final thought we observe that intuitively understandable systems probably have an advantage over less intuitive, more complex, and more mathematically advanced systems, as far as acceptance of the systems goes. It is our duty as researchers and model-builders to keep things as simple as possible, while still providing good-enough tools. Sometimes it is quite evident that over-modeling occurs and may create problems, even with the credibility of the results from the used systems. One remedy for these problems could be to re-read some classics that discuss our relationship with tools we create for management, such as "Management misinformation systems" by Russell Ackoff [1], a paper that is extremely up-to-date even today, almost fifty-years after its publication.

References

1. Ackoff, R.: Management misinformation systems. Manage. Sci. **14**(4), B147–B156 (1967)
2. Beliakov, G., Pradera, A., et al.: Aggregation functions: a guide for practitioners. Springer, Berlin (2007)
3. Bobillo, F., Delgado, M., et al.: A semantic fuzzy expert system for a fuzzy balanced scorecard. Expert Syst. Appl. **36**, 423–433 (2009)

4. Bojadziev, G., Bojadziev, M.: Fuzzy logic for business, finance, and management. World Scientific, Washington (2007)
5. Bradley, M., Desai, A., et al.: The rationale behind inter-firm tender offers: information or synergy? J. Applied Corporate Finance **16**, 63–76 (1983)
6. Bremser, W., Barsky, N.: Utilizing the balanced scorecard for R&D performance measurement. Res. Technol. Manage. **47**(6), 229–238 (2004)
7. Canfora, G., Troiano, L.: An extensive comparison between OWA and OFNWA Aggregation. In: 8th SIGEF Conference, Napoli, Italia (2001)
8. Carlsson, C., Fullér, R., et al.: A note on constrained OWA aggregation. Fuzzy Sets Syst. **139** (3), 543–546 (2003)
9. Chiclana, F., Herrera, F., et al.: Some induced ordered weighted averaging operators and their use for solving group decision making problems based on fuzzy preference relations. Eur. J. Oper. Res. **183**(1), 383–399 (2007)
10. Collan, M.: Fuzzy or linguistic input scorecard for IPR evaluation. J. Appl. Oper. Res. **5**(1), 22–29 (2013)
11. Collan, M., Kinnunen, J.: A procedure for the rapid pre-acquisition screening of target companies using the pay-off method for real option valuation. J. Real Options Strategy **4**(1), 115–139 (2011)
12. Collan, M., Kyläheiko, K.: Forward-looking valuation of strategic patent portfolios under structural uncertainty. J. Intellect. Property Rights **18**(3), 230–241 (2013)
13. Eilat, H., Golany, B., et al.: R&D project evaluation: an integrated DEA and balanced scorecard approach. Omega **36**(5), 895–912 (2008)
14. Filev, D., Yager, R.: On the issue of obtaining OWA operator weights. Fuzzy Sets Syst. **94**(2), 157–169 (1998)
15. Fullér, R., Majlender, P.: An analytic approach for obtaining maximal entropy OWA operator weights. Fuzzy Sets Syst. **124**, 53–57 (2001)
16. Fullér, R., Majlender, P.: On obtaining minimal variability OWA operator weights. Fuzzy Sets Syst. **136**(2), 203–215 (2003)
17. Hathout, I.: Damage assessment of existing transmission towers using fuzzy weighted averages. In: Second International Symposium on Uncertainty Modeling and Analysis College Park, MD, USA, IEEE (1993)
18. Heikkilä, M.: R&D investment decisions with real options—profitability and decision support. Department of Information Systems, Faculty of Technology Turku, Abo Akademi University. D. Sc. (Economics and Business Administration) (2009)
19. Herrera, F., Herrera-Viedma, E., et al.: Direct approach processes in group decision making using linguistic OWA operators. Fuzzy Sets Syst. **79**(2), 175–190 (1996)
20. Hitt, M.A., King, D., et al.: Mergers and acquisitions: overcoming pitfalls, building synergy, and creating value. Bus. Horiz. **52**, 523–529 (2009)
21. Kao, C., Liu, S.-T.: Competitiveness of manufacturing firms: an application of fuzzy weighted average. IEEE Trans. Syst. Man Cybern. Part A Syst. Hum. **29**(6), 661–667 (1999)
22. Kaufmann, M., Gupta, M.: Introduction to Fuzzy Arithmetics: Theory and Applications. Van Nostrand Reinhold, New York (1985)
23. Kaufmann, M., Gupta, M.: Fuzzy Mathematical Models in Engineering and Management Science, Elsevier Science Publishers B.V., Amsterdam(1988)
24. Li, G., Dalton, D.: Balanced Scorecard for R&D. Pharm. Executive. **23**, 84–90 (2003)
25. Luukka, P., Collan, M.: Fuzzy scorecards, FHOWA, and a new fuzzy similarity based ranking method for selection of human resources. In: IEEE conference on systems, man, and cybernetics (IEEE SMC 2013), Manchester, UK, IEEE (2013)
26. Merigo, J.M., Casanovas, M.: Geometric operators in decision making with minimization of regret. Int. J. Comput. Syst. Sci. Eng. **1**(1), 111–118 (2008)
27. Mitchell, H.B., Estrach, D.D.: An OWA operator with fuzzy ranks. Int. J. Intell. Syst. **13**, 69–81 (1998)

28. O'Hagan, M.: Aggregating template or rule antecedents in real time expert systems with fuzzy set logic. In: 22nd Annual IEEE Asilomar Conference on Signals, Systems, and Computers Pacific Grove, CA, USA (1988)
29. O'Hagan, M.: Fuzzy decision aids. In: 21st Asilomar Conference on signal, systems and Computers. Pacific Grove, CA, IEEE and Maple Press (1987)
30. Reilly, R.F., Schweihs, R.P.: Valuing Intangible Assets. McGraw-Hill Professional Publishing, Blacklick (1998)
31. Schaefer, P.A., Mitchell, H.B.: A generalized OWA operator. Int. J. Intell. Syst. **14**, 123–143 (2004)
32. Smith, G.V., Parr, R..: Valuation of Intellectual Property and Intangible Assets. Wiley, New York (2000)
33. Tee, A.B., Bowman, M.D.: Bridge condition assessment using fuzzy weighted averages. Civ. Eng. **8**(1), 49–57 (1991)
34. Torra, V., Narukawa, Y.: Modeling Decisions—Information Fusion and Aggregation Operators. Springer, Berlin (2007)
35. Uehara, K.: Fuzzy inference based on a weighted average of fuzzy sets and its learning algorithm for fuzzy exemplars. In: IEEE International Joint Conference on the Fuzzy Systems, IEEE (1995)
36. van den Broeck, P., Noppen, J.: Fuzzy weighted average: alternative approach. In: Annual Meeting of the North American Fuzzy Information Processing Society, NAFIPS, Montreal, Quebec, Canada, IEEE (2006)
37. Wang, Y.M., Luo, X.W., et al.: Two new model for determining OWA operator weights. Comput. Ind. Eng. **52**(2), 203–209 (2007)
38. Xu, Z.S.: Linguistic aggregation operators: an overview. In: Fuzzy Sets and Their Extensions: Representation, Aggregation and Models **220**(1), 163–181 (2008)
39. Xu, Z.S., Da, Q.L.: An overview of operators for aggregating information. Int. J. Intell. Syst. **18**(9), 953–969 (2003)
40. Yager, R.: On ordered weighted averaging operators in multicriteria decision making. IEEE Trans. Syst. Man Cybern. **18**(1), 183–190 (1988)
41. Yager, R.: Induced aggregation operators. Fuzzy Sets Syst. **137**(1), 59–69 (2003)
42. Yager, R.: Generalized OWA aggregation operators. Fuzzy Optim. Decis. Making **3**(1), 93–107 (2004)
43. Yager, R., Kacprzyk, J., et al.: Recent developments on the Ordered Weighted Averaging Operators: Theory and Practice. Springer, Berlin (2011)
44. Yager, R.R.: Heavy OWA operator. Fuzzy Optim. Decis. Making **1**(4), 379–397 (2002)
45. Zadeh, L.A.: Fuzzy Sets. Inf. Control **8**(1), 338–353 (1965)
46. Zadeh, L.A.: The concept of linguistic variable and its application to approximate reasoning. Inf. Sci. **1**(1), 199–249 (1975)
47. Zadeh, L.A.: Fuzzy sets as a basis for a theory of possibility. Fuzzy Sets Syst. **1**, 3–28 (1978)

Decision Analytics and Soft Computing with Industrial Partners: A Personal Retrospective

József Mezei and Matteo Brunelli

Abstract Methods in decision analytics are becoming essential tools for organizations to process the increasing amount of collected data. At the same time, these models should be capable of representing and utilizing the tacit knowledge of experts. In other words, companies require methods that can make use of imprecise information to deliver insights in real time. In this chapter, we provide a summary of three closely related research projects designed by building on the concept of knowledge mobilization. In these three cases, we provide solutions for typical business analytical problems originating mainly form the process industry. Fuzzy ontology represented as a fuzzy relation provides the basis for every application. By looking at the similarities among the three cases, we discuss the main lessons learnt and provide some important factors to be considered in future applications of soft computing in industrial applications.

1 Introduction

In a world of ever increasing complexity making informed decisions becomes a more and more demanding task. The old problem of not having enough data has often become the opposite problem of having an over-abundance of data. With such an increasing amount of data, new challenges emerge. Among them, there are the following three: (i) find and isolate relevant data in the ocean of all data (finding the needle in the haystack), (ii) transform data into information and ultimately knowledge, (iii) capture the tacit knowledge of experts into a structured form and make use of it. In this context, and facing these problems, we have been involved in three

---

J. Mezei (✉)
Department of Information Technologies, Åbo Akademi University, Turku, Finland
e-mail: jmezei@abo.fi

M. Brunelli
SAL, Department of Mathematics and Systems Analysis, Aalto University, Espoo, Finland
e-mail: matteo.brunelli@aalto.fi

M. Collan et al. (eds.), *Fuzzy Technology*, Studies in Fuzziness and Soft Computing 335, DOI 10.1007/978-3-319-26986-3_11

consecutive research projects in cooperation with the Finnish Funding Agency for Technology and Innovation (Tekes) and various industrial partners.

In this book chapter we would like to offer an account of how we used techniques borrowed from soft computing (mainly fuzzy set theory) to build analytical tools in cooperation with industrial partners. The remaining of the chapter is organized as follows. In the next section we discuss the emerging role of decision analytic tools. In Sects. 3, 4, and 5 we discuss the problems posed by industrial companies and briefly describe the proposed solutions to their needs. Each section corresponds to a different research project. Furthermore, for sake of relevance, in each section, we decided to spell out, in an *ad hoc* created environment, the real-world research question coming from the industrial partners. In Sect. 6, on the ground of the experience gained in the aforementioned projects, we reflect on our experiences of research in cooperation with industrial partners bringing real-world research problems. In Sect. 7 we draw some conclusions.

2 Decision Analytics and Soft Computing

As more data is available than ever before, the main task faced by industrial decision makers is to transform the massive amount of information into insights that create value and offer an advantage over competitors [15]. In contrast to previous decades in which many organizations were lacking sufficient data, the situation is transformed into the case when the organizations have more data than they can use effectively. Decision analytics emerged in the previous decade as a collection of analytical tools that can provide an efficient mean to cope with different data-sets.

A formal definition of analytics was given by Davenport and Harris (2007, p. 7) [6] as "the extensive use of data, statistical and quantitative analysis, explanatory and predictive models, and fact-based management to drive decisions and actions". Recent surveys indicate that more than 60 % of organizational decisions are based on analytic inputs and most of the managers believe that they will need to increase their analytic resources [16].

Based on these positive developments, the rapid evolution of analytic tools, and the processing power of computers, one would expect that more and more companies implement analytics solutions to leverage the potential benefits. On the other hand, according to research by the analyst firm Gartner [7], 70–80 % of corporate analytics projects fail and, according to a survey focusing on specifically big data, 55 % of big data projects do not get completed and many others fall short of their objectives.

The main context of our present research is the manufacturing industry which is the second most important sector of the Finnish economy, after services, and the key sector considering foreign trades (16.7 % of the GDP in 2013 [19]). The key branches [19] include pulp and paper industry (8.6 %), machinery and equipment (13.1 %), and chemicals and chemical products (4.8 %). There exist numerous applications of different analytic techniques in process industry applications: statistical methods (for example Bayesian predictions [13]), artificial neural networks (for process emission monitoring [10]), or optimization (process industry supply chains [1]).

In these industries, with the rapid development of sensor technology, the amount of data collected from complex machines and equipment is even higher than in other contexts [11], and they become essential to effective operations. On the other hand, a different type of information, namely expert (tacit) knowledge, is available and used on the individual level in daily operational decision making problems. This type of information is difficult to capture and utilize in decision making problems. In theory, one of the most crucial differences between analytics and its predecessors lies in the role of experts: they are active participants from the development to the deployment processes. In practice, as a consequence of the difficulties related to making use of expert knowledge, this aspect is usually neglected. This can be one of the possible reasons for the failure of analytics projects: companies focus mainly on the technology and methods rather than considering the attributes of the problem at hand by capturing the tacit knowledge of experts. The developed models in theory are ideal, but the development process loses sight of the practical contexts of the projects which is mainly a consequence of the lack of communication between data people and decision people [8].

Our proposal as one possible solution for this problem is to combine analytics technologies, specifically soft computing, with task-specific knowledge in order to transform data into insights and insights into value. We will describe the features of the general process that we followed in developing analytical tools for managing knowledge through three cases. The common starting point and problem for all the cases is that in many practical situations, experts cannot describe their knowledge in precise terms or formulate their knowledge in a directly reusable way for future applications. To overcome this, we propose to use soft computing methodologies, more specifically fuzzy logic, to capture, represent, and make use of expert knowledge. We found that fuzzy ontologies, which in our case are represented by fuzzy relations, provide a tool that can be appropriate to solve the above mentioned problems.

There exists several soft computing methodologies that are closely connected to analytics and appear frequently in analytics related literature, such as neural networks and metaheuristic optimization methods. Although we can also observe a slightly increasing presence of fuzzy logic based models, frequently it is still not mentioned as an important methodology. In this paper, we show the usefulness of fuzzy logic in problems that are typical in decision analytics and are approached from the point of view of expert knowledge. An important feature of the projects described in the following sections is that they are the outcome of university-industry cooperation. We will also shortly reflect on the best practices proposed by Pertuze et al. [18] for these types of collaboration and the main differences compared to cases when everybody involved in a project comes from the same organization.

3 Knowledge Mobilisation (2008–2010)

One of the foremost problems in industrial processes is that of choosing among different alternatives and courses of action. In many real world situations, alternatives and courses of action can be described by means of their characteristics. One example of set of alternatives is that of wines, since each wine can be described by means of a large number of characteristics as, for example, 'percentage of alcohol', 'acidity', 'color', and so forth. The situation described by wines can be transposed to real industrial problems where, for instance, different chemicals used in production processes can be described by their chemical properties like 'pH', 'melting temperature', 'solvability', and many others. Hence, although for sake of clarity here we shall speak of wines, the reader should bear in mind that the underlying idea can be extended to be used with other alternatives.

It is in this context that an expert might want to inquiry a database of wines and ask for a 'mildly alcoholic' wine which goes well with 'chicken' or 'beef'. From the basic assumption that queries can be stated in logical form using only *and*, *or* and *complement* operators, the following research question was posed to us when cooperating with industrial partners.

Question 1 Given the description of alternatives, e.g. wines or chemicals, by means of their characteristics, how can we select the most suitable alternative to a logical query?

Before the database can be used to answer users' queries, it is fundamental to populate and model it in a compatible way. The first step for the formal representation of the database of a set of wines was the creation of a fuzzy ontology which could capture their characteristics in an homogeneous form. The fuzzy ontology was expressed in the form of a fuzzy relation R associated to the membership function

$$\mu_R : A \times C \to [0, 1]$$

where A and C are the sets of alternatives and their characteristics, respectively. For convenience in the modeling of the query, the set C comprehends different levels for the characteristics. For instance, C does not include 'alcoholic' but, instead, 'low alcohol', 'medium alcohol', and 'high alcohol'. This also allows the user to express his queries in a linguistic way. Table 1 contains an excerpt from the fuzzy ontology of wines.

One of the most delicate steps was the definition of the membership function μ_R. Three methods were used to estimate it:

- Some values of the relation were given subjectively by experts. For instance, in the case of wines, the experts might state how well a wine fits with a certain type of occasion.
- When the characteristic can be expressed on a numerical scale, fuzzy numbers can be used to estimate the relation between a wine and a given characteristic. Consider, for example, the relative amount of alcohol in the wine. The value of

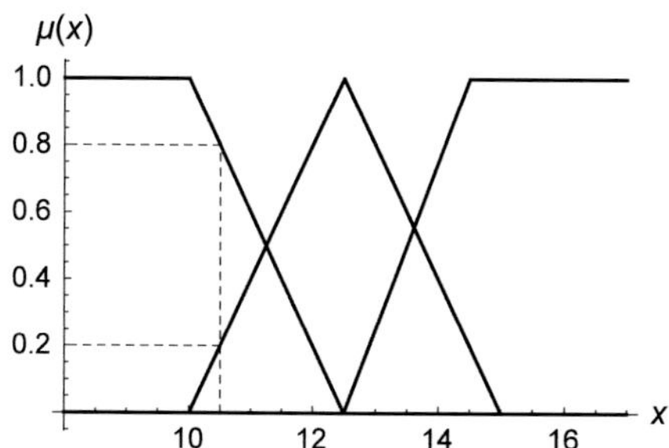

Fig. 1 Representations of the membership functions of the sets 'low alcohol', 'medium alcohol', and 'high alcohol' where x is the alcohol percentage. According to the figure, a wine with the 10.5 % of alcohol belongs to the set of low alcoholic and medium alcoholic wines with degrees 0.8 and 0.2, respectively

the membership function of a given wine can be established thanks to some fuzzy numbers. The fuzzy numbers for 'low alcohol', 'medium alcohol', and 'high alcohol' are plotted in Fig. 1.

- Rules can be defined to estimate unknown values of the relation starting from known ones. For example, the value of the relation between a given wine and the type of food to be served with could be a function of the color of the wine, its acidity, its alcoholic level and so on.

Once the ontology is complete we have a formal representation of the knowledge on the wines, which ranges from their acidity to their suitability to certain occasions or certain drinkers. The second phase is therefore that of exploitation of this knowledge base to make decisions. As mentioned before, when choosing the most suitable wine, it is customary to ask logical queries as, for instance "I would like a wine which goes well for chicken *or* beef, *and* to be enjoyed with friends, but it should *not* be highly alcoholic". Also in this case, fuzzy sets theory provides tools to solve this problem.

Triangular norms (t-norms), triangular conorms (t-conorms), and fuzzy complement operations have been used in fuzzy set theory to generalize the operations of intersection, union and complement, respectively [14]. One could consider the product operation $(a \cdot b)$ to be used as t-norm and its dual t-conorm $(a + b - a \cdot b)$ to model *and* and *or* logical connectors, respectively. Consider the query of someone seeking for "a wine with medium alcohol *and* suitable for fish *or* chicken". Then, this query, which we can call Q, defines a fuzzy subset of the set of all wines. Referring to Table 1, the membership value of the various wines in Q is calculated as follows:

$$\begin{aligned}
\mu_Q(\text{Tommasi Crearo}) &= 0.8 \cdot (0.43 + 0.75 - 0.43 \cdot 0.75) = 0.686\\
\mu_Q(\text{Trimbach}) &= 0.8 \cdot (0.66 + 0.08 - 0.66 \cdot 0.08) = 0.54976\\
\mu_Q(\text{Morada}) &= 0.5 \cdot (0.37 + 0.5 - 0.37 \cdot 0.5) = 0.3425\\
\cdots &= \cdots\\
\mu_Q(\text{El Tiempo}) &= 0.6 \cdot (0.49 + 0.1 - 0.49 \cdot 0.1) = 0.2976
\end{aligned}$$

Table 1 Excerpt from the *fuzzy* wine ontology [3]

	Country			Alcohol			Food			…	Drinkers		
	Italy	…	France	Low	Medium	High	Fish	…	Pork	…	Connoisseur	…	Pretender
Tommasi Crearo	1	…	0	0	0.80	0.25	0.43	…	0.75	…	0.38	…	0.81
Trimbach Pinot Gris Reserve	0	…	1	0	0.80	0.25	0.66	…	0.08	…	0.29	…	0.85
Morada Aged Tempranillo	0	…	0	0	0.5	0.6	0.37	…	0.5	…	1	…	0.5
…	…	…	…	…	…	…	…	…	…	…	…	…	…
El Tiempo Rosado	0	…	1	0.4	0.6	0	0.49	…	0.1	…	0	…	0.8

Hence, Tomasi Crearo is the wine which best fits the query Q and shall be recommended to the end user of the system. The name of the project, Knowledge Mobilization, evokes the fact that knowledge shall not only exist, but also be readily available. Hence, part of the project involved the realization of an online platform for decision making based on fuzzy ontology [3]. Besides the research described in the next two sections, Knowledge Mobilisation has inspired some other extensions [17].

4 Dyscotec (2010–2012)

Another problem faced by industries operating large and complex machines is to understand the possible causes of failure of the machines and having a prompt suggestion for their solution. In fact, although industries accept that, in spite of preventive maintenance, from time to time machines can fail, they want to keep the length of the disruption of the production line as short as possible.

Often, when problems are encountered and machines are very complex, e.g. industrial paper machines, technicians are required to write short reports indicating the causes of the failure and a description of how they eventually solved the problem at issue. Usually, these reports are written in a textual form, each of them is associated to some keywords, and they are all electronically stored in a database. Such reports become useful when *new* technicians face *old* problems.

The answer to the following question can help technicians of industrial machines to retrieve a report where perhaps the problem and its solution were already described.

Question 2 Given a database of past reports, how can we retrieve the report which most likely contains the solution of the problem?

First of all, each report is associated with a list of keywords. Therefore, the database can be described as a 0-1 table where the cell (i, j) has value 1 if the ith document (d_i) contains the jth keyword (k_j), and 0 otherwise. Table 2 is a toy example of a database with only three documents and four keywords.

One problem related with the search with keywords is that they cannot be considered as strings of letters. If they were considered so, then the word ‘car’ would appear very similar to ‘cat’, and very dissimilar to ‘automobile’, although they are synonyms. On the contrary, it is auspicable that, if the expert is looking for ‘car’, the reports containing ‘automobile’, and not ‘cat’, be retrieved by the system.

Table 2 Excerpt from a report database

	k_1	k_2	k_3	k_4
d_1	0	1	0	1
d_2	0	0	1	0
d_3	1	0	0	1

For example, document 1 contains keywords 2 and 4

Considering the set of possible keywords (or a subset of them), which we call $K = \{k_1, \ldots, k_n\}$, we proposed [5] to build a fuzzy relation with membership function $\mu_S : K \times K \to \mathscr{F}[0,1]$ where $\mathscr{F}$ is the set of trapezoidal fuzzy numbers with support in $[0, 1]$. The fuzzy number $\mu_S(k_i, k_j)$ reflects the *degree of semantic similarity* of the keyword k_i to the keyword k_j. That is, the greater the fuzzy number, the more similar the keywords. The following is an example of fuzzy relation for the keywords in Table 2. Note that the similarity between a keyword and itself is always equal to 1.

$$\mathbf{S} = \begin{pmatrix} (1,1,1,1) & (0,0,0.25,0.5) & (0.5,0.5,0.5,0.5) & (0,0.25,0.5,0.75) \\ (0,0,0.25,0.5) & (1,1,1,1) & (0.2,0.2,0.2,0.2) & (0.2,0.2,0.3,0.3) \\ (0.4,0.4,0.4,0.4) & (0.3,0.3,0.5,0.5) & (1,1,1,1) & (0.6,0.6,0.8,0.8) \\ (0.3,0.3,0.3,0.3) & (0.1,0.1,0.4,0.4) & (0.25,0.5,0.75,1) & (1,1,1,1) \end{pmatrix} \tag{1}$$

Trapezoidal fuzzy numbers have been chosen to represent degrees of similarity since they are more general than real numbers and real intervals and they have been often used to represent linguistic expressions; in this sense they provide a general enough representation.

Now consider the two keywords ‘car’ and ‘automobile’. If the document *explicitly* contains ‘car’ as a keyword, but not ‘automobile’, then it is safe to assume that, when it comes to the meaning of the document, probably it *implicitly* contains also the keyword ‘automobile’, to the extent to which this is similar to ‘car’.

Our strategy to account for the fact that similar concepts can be expressed by means of different words is that of replacing the generic original value for the pair (i,j) in Table 2 with the greatest value of similarity to any other keyword present in the document. Consider, for instance, the cell $(1,3)$ in Table 2. It has value 0 since d_1 does not contain k_3. However, d_1 contains some keywords which, to some extent, are similar to k_3. Hence, we replace the value 0 with the greatest among the degrees of similarity of k_3 with the keywords used in the document, i.e. $\max\{\mu_S(k_3, k_2), \mu_S(k_3, k_4)\}$, which is $\mu_S(k_3, k_4) = (0.6, 0.6, 0.8, 0.8)$.

At this stage, we have a set of keywords as an input from the expert who is seeking for helpful reports. Such a set can be represented as a binary vector of length n, where n is the number of keywords. The ith component of the vector is equal to 1 if and only if the keyword is searched for, and 0 otherwise. The real numbers 0 and 1 can be interpreted as special cases of fuzzy numbers and therefore distance measures between fuzzy numbers can be used to rank the reports—which are here described as the rows of Table 3—from the most to the least likely to include those keywords (or their synonyms).

Table 3 Excerpt from a report database

	k_1	k_2	k_3	k_4
d_1	$(0, 0.25, 0.5, 0.75)$	$(1, 1, 1, 1)$	$(0.6, 0.6, 0.8, 0.8)$	$(1, 1, 1, 1)$
d_2	$(0.5, 0.5, 0.5, 0.5)$	$(0.2, 0.2, 0.2, 0.2)$	$(1, 1, 1, 1)$	$(0.25, 0.5, 0.75, 1)$
d_3	$(1, 1, 1, 1)$	$(0, 0, 0.25, 0.5)$	$(0.6, 0.6, 0.8, 0.8)$	$(1, 1, 1, 1)$

5 Data to Intelligence (D2I, 2012–)

Continuing the work done in the previous two projects, we employed the fuzzy ontology (fuzzy relation) to tackle a general situation: optimizing the operations of a complex machine modeled in terms of input and output variables. Specifically, the main motivation was to understand the processes taking place in a paper machine by making use of the knowledge of experts who have been working with the machines for decades [4]. As a result of mainly market-related issues, e.g. the continuously decreasing demand for paper products globally, companies in the paper industry started to hire young engineers without extensive work experience with paper machines. To cope with this situation, companies aim at creating decision support tools that can partially automate the calibration of paper machines by capturing the knowledge of retiring, knowledgeable experts. Hence, the following was the question that we were asked to answer.

Question 3 How can we reuse the expertise of retiring engineers and help the new engineers to calibrate industrial paper machines?

As a starting point of creating a tool for this purpose, the expert is asked to represent his/her knowledge in a way that s/he thinks is the most appropriate as a representation. The paper machine expert involved in the project described his knowledge in terms of a heat map: a graphical representation of the relationship between an identified set of inputs to the paper machine, termed as *factors*, and a set of output indicators of the produced paper, termed as *characteristics*. In practice, an engineer can change the value of a characteristic of a paper to a required level by changing the value of a subset of the factors. A change in a factor can positively or negatively effect a characteristic or in some cases it does not have any effect. The intensity of this effect is indicated in the heat map by different shades of colors. As in the previous two applications, also in this case a fuzzy relation can provide a quantitative representation of the relationships between the set of factors $F = \{f_1, \dots, f_n\}$ and the set of characteristics $C = \{c_1, \dots, c_m\}$ as:

$$\mu_R(f_i, c_j) = \begin{cases} 1, & \text{if } f_i \text{ strongly positively affects } c_j \\ \alpha \in]0, 1[, & \text{if } f_i \text{ to some extent positively affects } c_j \\ 0, & \text{if } f_i \text{ does not affect } c_j \\ \beta \in]-1, 0[, & \text{if } f_i \text{ to some extent negatively affects } c_j \\ -1, & \text{if } f_i \text{ strongly negatively affects } c_j. \end{cases} \tag{2}$$

By relying on this relation as a representation, different optimization models can be proposed for "tuning" a paper machine. In practical problems, it is hardly ever the case that there is a single characteristic of the paper to be changed; an improvement in the quality of the paper usually requires several characteristics to be optimized at the same time. The general task consists of specifying a partition of the characteristics (to be decreased, increased, left unaltered or do not care) and finding a partition of factors (to increase, decrease or do not change) that corresponds to each other to an acceptable extent based on the fuzzy relation.

The basic optimization model can be extended in different ways to provide a more accurate and reliable presentation of the context and a more customizable tool for the intended users, the engineers with limited practical experience with the paper machines. The first improvement concerns the imprecision present in the color selection of the heat map and consequently in the values of the fuzzy relation. As an attempt to overcome this problem, the real number indicating the relationship between a factor and characteristic can be represented using a (triangular) fuzzy number. This would also allow for using the possibilistic version of chance constrained programming in the optimization model: according to the preferences of the paper machine engineer, an acceptable degree of satisfaction regarding the partition of characteristics can be defined in the form of a possibilistic constraint. For example, we accept a solution as satisfactory if the possibility that the identified setting of factors increases the brightness of the paper is at least 0.9.

A different extension aims at incorporating additional information about paper machines. This includes, for example, the cost of changing the value of different factors. As companies have limited resources in terms of time and money, it is very important to incorporate this aspect of the decision process in the optimization model to obtain a reasonable solution, not only a theoretically correct one. We can also refine the partitions based on detailed empirical information of the range of the values of the factors and how the relationship with a given characteristic may vary in different intervals of the range.

6 Discussion

In this section, we will summarize the lessons learnt during the research process in the three discussed cases. In our opinion, the most important points are the following ones.

- We demonstrated how to make use of expert knowledge captured in three different ways (rules, reports, heat-map). Our approach reflected several of the points raised by Kaisler et al. [12] concerning advanced analytics. Most importantly, from the beginning the goal was to create knowledge-centric systems with the help of the knowledge mobilization approach. The main features of this approach include the following characteristics: (i) the system should be user-centric, regardless on whether it is about mobile phone users in a restaurant choosing a wine or engineers tuning a paper machine; (ii) it should be context-adaptive and be easy to fit the model to the problem description automatically; and (iii) it should perform smart operations in the sense that the reasoning behind the models tries to approximate the reasoning of human experts.

- All the different representations can be formulated as a fuzzy ontology/fuzzy relation. There exists several definitions of fuzzy ontology and numerous theoretical contributions with only a very few practical case studies and applications. Bobillo [2] defined fuzzy ontology as "an ontology which uses fuzzy logic to provide a natural representation of imprecise and vague knowledge and eases reasoning over it". This definition does not stress any necessary condition on the complexity of the reasoning system; accordingly, for us, the main goal was to use a concept that can be easily operationalized. We found that a representation as a fuzzy relation provides sufficient rigor and at the same time helps in creating applications that solve important real-life problems.
- The problems that we approached belong to the core of decision analytics. According to the classification of Holsapple et al. [9], we considered developing and implementing analytics as a transformation process to drive organizational decisions, in our examples on the operational level. The knowledge of experts with many years of experience is a crucial success factor of the transformation supported by analytics. In the discussed cases, we illustrated how soft computing methods, specifically fuzzy logic, can be seen as an essential component of an analytic transformation by providing a tool to capture, represent and utilize tacit knowledge.
- Prescriptive analytics cases are not documented in large numbers in the literature; here we illustrated three cases. Analytics methods are usually classified into three main groups: descriptive, predictive, and prescriptive. Descriptive and predictive methods are widely used and present in every organization and used on a daily basis as they provide answers for the question what happened (and why) and what will happen in the future, respectively. Prescriptive analytics, the most advanced class of analytics, is not only concerned with estimating possible future events but also at the same time tries to identify the best possible course of action (decision) to react to predicted events. Although this approach could offer the most potential advantages for organizations, presently only 3 % of companies employ prescriptive analytics. From this perspective we contributed to research by describing the process of designing and creating prescriptive analytics systems.

Additionally, the 5-years long process of working on these problems continuously in close cooperation with industrial partners supports most of the observations described by Pertuze et al. [18]. In our opinion, the first of several main reasons for the success was that the approached problems are rooted in and motivated by actual problems of the companies that are necessary to be solved in order to keep up the existing effectiveness in different processes when facing the rapidly changing global market and continuously changing customers' needs. Secondly, the duration of the research projects, and the way the three cases build on each other, allowed for continuous improvements in the methodologies while at the same time frequently reflecting back on the original concepts. Thirdly, the extensive involvement of company experts in improving the models and providing the contextual knowledge resulted in insights that would have been hardly achievable without the use of domain knowledge.

7 Conclusions

In this chapter, we recalled three closely connected research projects taking place between 2008 and 2014 as a collaboration between industrial companies, funding agencies, and University. All the three described cases involved analytical methods and aimed at capturing expert knowledge and creating new insights. This general problem description classifies our cases as belonging to the problems of decision analytics. Based on the experience learnt in the three projects, we highlight the commonalities that we think should be taken into consideration when developing analytics research projects in the future. We illustrate the important role that soft computing, and specifically fuzzy logic, can play in creating prescriptive analytics solutions.

Every case contributes to the general concept of knowledge mobilization: creating advanced analytical tools that can provide context adaptive knowledge to the end-users in real time. In the described cases, the imprecision of the experts' knowledge and their linguistic expressions in different reports were modeled using fuzzy ontologies/relations. Fuzzy relations/ontologies provided an appropriate tool to capture and manipulate imprecise information in the different industrial contexts.

Acknowledgments This paper is dedicated to the memory of Péter Majlender whose important contributions to fuzzy set theory have influenced us. József Mezei acknowledges the support from the TEKES strategic research project Data to Intelligence [D2I], project number: 340/12. The research of Matteo Brunelli is supported by the Academy of Finland.

References

1. Barbosa-Póvoa, A.P.: Progresses and challenges in process industry supply chains optimization. Curr. Opin. Chem. Eng. **1**(4), 446–452 (2012)
2. Bobillo, F.: Managing vagueness in ontologies. Ph.D. thesis, University of Granada, Spain (2008)
3. Carlsson, C., Brunelli, M., Mezei, J.: Decision making with a fuzzy ontology. Soft Comput. **16**(7), 1143–1152 (2012)
4. Carlsson, C., Brunelli, M., Mezei, J.: A soft computing approach to mastering paper machines. In: Proceedings of the 46th Hawaii International Conference on System Sciences (HICSS), pp. 1394–1401. IEEE (2013)
5. Carlsson, C., Mezei, J., Brunelli, M.: Fuzzy ontology used for knowledge mobilization. Int. J. Intell. Syst. **28**(1), 52–71 (2013)
6. Davenport, T.H., Harris, J.G.: Competing on analytics: The new science of winning. Harvard Business Press, Boston (2007)
7. Goodwin, B.: Poor Communication to Blame for Business Intelligence Failure, Says Gartner. http://www.computerweekly.com/news/1280094776/Poor-communication-toblame-for-business-intelligence-failure-says-Gartner (2011). Accessed 7 Oct 2014
8. Guszcza, J., Lucker, J.: Why Some CEOs Are So Skeptical of Analytics?. http://deloitte.wsj.com/cio/2012/06/05/403/ (2012). Accessed 7 Oct 2014
9. Holsapple, C., Lee-Post, A., Pakath, R.: A unified foundation for business analytics. Decis. Support Syst. **64**, 130–141 (2014)

10. Iliyas, S.A., Elshafei, M., Habib, M.A., Adeniran, A.A.: RBF neural network inferential sensor for process emission monitoring. Control Eng. Pract. **21**(7), 962–970 (2013)
11. Kadlec, P., Gabrys, B., Strandt, S.: Data-driven soft sensors in the process industry. Comput. Chem. Eng. **33**(4), 795–814 (2009)
12. Kaisler, S.H., Espinosa, J.A., Armour, F., Money, W.H.: Advanced analytics–issues and challenges in a global environment. In: Proceedings of the 47th Hawaii International Conference on System Sciences (HICSS), pp. 729–738. IEEE (2014)
13. Khatibisepehr, S., Huang, B., Khare, S.: Design of inferential sensors in the process industry: a review of bayesian methods. J. Process Control **23**(10), 1575–1596 (2013)
14. Klir, G., Yuan, B.: Fuzzy Sets and Fuzzy Logic: Theory and Applications. Prentice Hall, New Jersey (1995)
15. LaValle, S., Lesser, E., Shockley, R., Hopkins, M.S., Kruschwitz, N.: Big data, analytics and the path from insights to value. MIT Sloan Manage. Rev. **21**, 20–31 (2013)
16. Liberatore, M.J., Luo, W.: The analytics movement: implications for operations research. Interfaces **40**(4), 313–324 (2010)
17. Pérez, I.J., Wikström, R., Mezei, J., Carlsson, C., Herrera-Viedma, E.: A new consensus model for group decision making using fuzzy ontology. Soft Comput. **17**(9), 1617–1627 (2013)
18. Pertuze, J.A., Calder, E.S., Greitzer, E.M., Lucas, W.A.: Best practices for industry-university collaboration. MIT Sloan Manage. Rev. **51**, 83–90 (2010)
19. Statistics Finland: Finland in figures—national accounts. http://www.stat.fi/tup/suoluk/suoluk_kansantalous_en.html (2013). Accessed 7 Oct 2014

Spatial Analysis Using GIS for Obtaining Optimal Locations for Solar Farms—A Case Study: The Northwest of the Region of Murcia

J.M. Sánchez-Lozano, M.S. García-Cascales, M.T. Lamata and J.L. Verdegay

Abstract One of the first decisions that must be taken when hosting a photovoltaic solar farm to pour the energy generated into the grid is to choose a proper location (towns, existing infrastructure, etc.). The legislative framework that is applicable must also be considered since it involves a large number of restrictions (protected areas, streams and watercourses, etc.) that will provide us with the guidelines to eliminate those unsuitable areas, as well as certain criteria (proximity to power lines, slope, solar irradiation, etc.) according to which an evaluation of the suitable areas that condition any facility will be made. It is precisely for these reasons why the management of spatial visualization tools such as Geographic Information Systems (GIS) is particularly useful. The objective of this paper is to demonstrate how the aggregation of GIS to decision procedures in the field of renewable energy can solve complex location problems. In the present case a GIS (called gvSIG) will be employed in order to obtain suitable locations to host photovoltaic solar farms in the Northwest of the Region of Murcia, in Spain.

Keywords Geographic information systems · Solar farms · Decision making

J.M. Sánchez-Lozano
Centro Universitario de La Defensa. Academia General Del Aire,
Universidad Politécnica de Cartagena, Murcia, Spain

M.S. García-Cascales (✉)
Dpto. de Electrónica, Tecnología de Computadoras Y Proyectos,
Universidad de Politécnica de Cartagena, Murcia, Spain
e-mail: socorro.garcia@upct.es

M.T. Lamata · J.L. Verdegay
Dpto. Ciencias de La Computación E Inteligencia Artificial,
Universidad de Granada, Granada, Spain

M. Collan et al. (eds.), *Fuzzy Technology*, Studies in Fuzziness and Soft Computing 335, DOI 10.1007/978-3-319-26986-3_12

1 Introduction

More and more frequently, the Earth is warning us of the dangers involved in carrying out uncontrolled and disproportionate industrial development. The planet's response is manifested in the form of an increased number of forest fires, rising levels of seas and oceans, more droughts, extreme storms, as well as constant and frequent heat waves. These are just some of the most significant effects that are being caused by increased temperatures on our planet as a result of the increase in emissions of greenhouse gases [1]. From an energy point of view, in addition to the problem of global warming we must add a continuous rise in the prices of fossil fuels such as petroleum and natural gas [2]. Therefore, the human being must find alternatives that take advantage of the multitude of resources available and which will mitigate these significant negative effects.

In the last century, policies were developed globally [3–7] and at European level [8, 9] which promoted sustainable development strategies [10] and encouraged the implementation of renewable energy (RE) installations, with the objective that such technologies should play an important role in the generation of electrical energy in the future (Fig. 1).

In Spain, the fulfillment of the objectives set by the European Union was the main reason why different energy plans were developed [11, 12]. The objective was to reach at least 20 % of final energy consumption for 2020 using renewable technologies. As a result of this favourable legislative framework, in Spain the implantation of this type of facility was extended, with solar photovoltaics being one of those with the highest growth. This positioned Spain as the second world photovoltaic power in 2009 [13]. Recent analyses [14] have demonstrated that solar technologies, and in particular photovoltaic technology, have a stable learning curve, allowing to reach very high yields in regions where there is high solar radiation.

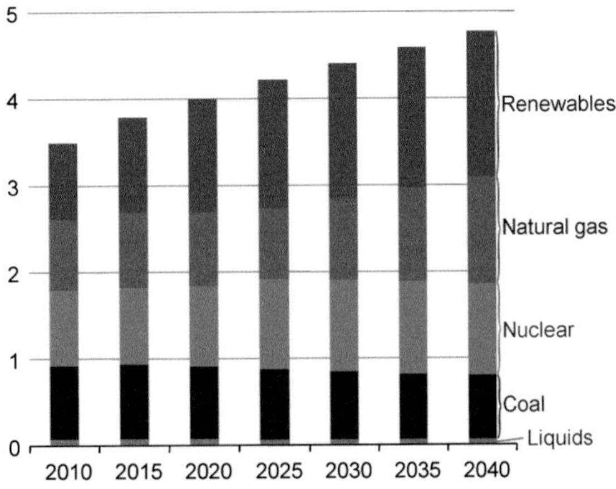

Fig. 1 Forecast for the generation of electrical energy by fuel in the European countries belonging to OECD, (trillions kWh) [2]

The objective of this work is to find the best location to install a solar photovoltaic farm. To do this, it will be necessary to find an area where solar radiation is high. On the other hand, it is necessary to define a number of attributes (slope, area, field orientation, distance to main roads, distance to power lines, distance to towns or villages, distance to electricity transformer substations, solar irradiation and average temperature) that will be important to evaluate the most favourable suitable location for the implementation of the solar farm.

The Region of Murcia is situated in the southeast of Spain, and has one of the highest levels of solar radiation in the country [15]. Therefore, it has become one of the areas with greater appeal to deploy photovoltaic solar farms. However, on its territory there are inland areas (Northwest region) which for various reasons (land less appropriate for the development of agriculture, low land prices, lower urban and residential occupation, etc.) present greater suitability than others to implement such facilities. So, an in-depth analysis, enabling to locate the best areas to deploy photovoltaic solar farms is of notable interest.

To carry out studies of this nature, it is evident that management tools such as geographic information systems (GIS) are very useful [16] since they are able to provide extensive databases, in the form of thematic layers and tables, which can be very useful to solve complex location problems [17].

2 GIS Methodology in Obtaining Suitable Surfaces to Deploy Installations of RE in the Northwest of the Region of Murcia

GIS are tools that manage geo referenced information and allow us to digitally represent the real world based on discrete objects. The information of these objects is expressed numerically and provides a collection of referenced data that acts as a reality model. The space data in a GIS is a set of maps that represent a portion of the actual surface, so that each one of these maps is defined by means of a thematic variable and when it is introduced in a GIS it receives the name of thematic layer.

2.1 The gvSIG Software

Although commercial GIS are widespread nowadays (ArcGIS, IDRISI, etc.), this paper will use an open source version called gvSIG (www.gvSIG.org); this was developed in 2004 by the Ministry of infrastructure and transport of the Valencian Community and is available to any user for its use and development.

In the proposed study, thematic layers that represent and define the surface covering the study area will be introduced in gvSIG, as will the restrictions i.e., areas in which it is already impossible to implement photovoltaic solar farms, because the current state of the terrain prevents it or it is prohibited by the legislation in force. Through the operations of edition of gvSIG, it will be possible to reduce the initial zone of study, taking into account the restrictions that affect it until the locations that are feasible to implement this type of facility are obtained.

2.2 *Stage 1: Search for Viable Locations*

The first stage will consist in selecting and refining the study area. In the proposed problem the zone will correspond to the Northwest of the Region of Murcia, which consists of five municipalities (Moratalla, Caravaca de la Cruz, Bullas, Calasparra, and Cehegin) and has an area of 2,379.62 km^2 (Fig. 2). Once the area of study is known, then the restrictions to apply will be described. These are the areas in which due to the current status of the territory (roads, railway, urban lands, etc.) and legislation, (European, national and regional regulations) it is not possible to deploy photovoltaic solar farms.

Each one of the seven constraints (Table 1) will be defined on the basis of the legislative framework that may apply so that, according to the current regulations

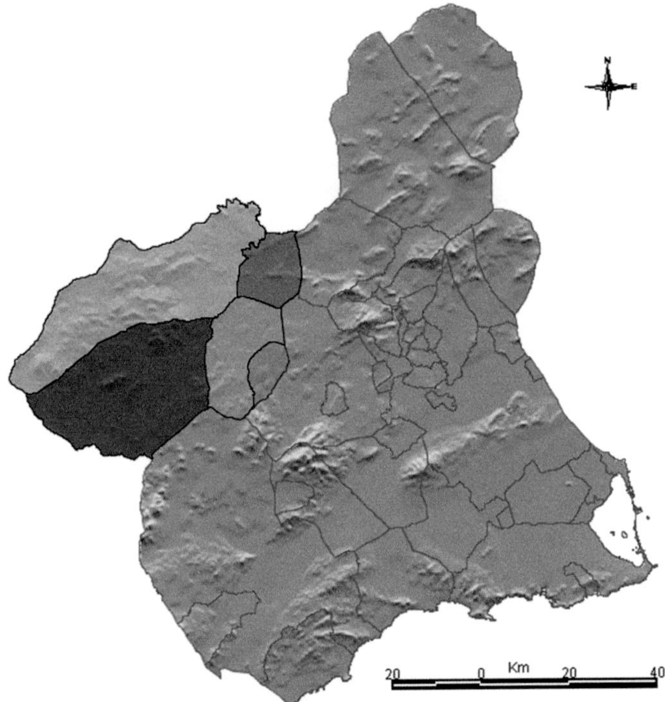

Fig. 2 Northwest of Murcia region

Table 1 Legal restrictions

Nº	Name of restrictions
1	Suitable for urban development and urban lands
2	Landscape value, water infrastructure, military areas and cattle trails
3	Runways and watercourses and streams
4	Archaeological paleontological and cultural heritage
5	Road and railway network
6	Sites of community importance (LICs)
7	Special protection areas for birds (ZEPAs)

[18–21], photovoltaic solar farms may not be implanted in any urban lands or lands suitable for urban development (restriction 1).

According to the law 42/2007 and law 3/1995 of 23 March, areas of high landscape value and earmarked for water supply infrastructures, military zones and cattle trails (restriction 2) are also protected areas. In Runways and watercourses and streams (restriction 3) and in their bands of buffer it is not possible to implant an installation [19]. In addition, law 16/1985 of 25 June, and the Legislative Decree 1/2005, establishes measures for the conservation of scheduled areas such as archaeological, paleontological and cultural heritage (restriction 4). The road and rail networks (restriction 5) are also protected by the regulations in force [24], as well as places of LICs (restriction 6) and areas of special protection for birds (restriction 7) which are protected by the Directive 92/43/EEC of 21 May 2009 [25].

In addition to the above restrictions, we must consider a further two factors, since those areas having any construction or installation of importance in its interior (marsh, agricultural construction, etc.), or that its area is less than that which experts consider to be the minimum to implement this type of installations (less than 1000 m^2 surface area) will be necessary to be discarded. Each of the restrictions will be introduced in the software gvSIG in the form of thematic layers. The thematic layer of the Northwest of the Region of Murcia will be initially introduced in gvSIG (Fig. 2). This layer will not only serve to delimit the study area but also to classify it by means of municipalities (which in turn divide their territory into polygons, plots and cadaster subplots). The thematic layers of restrictions will subsequently be added in a way which, with the commands of gvSIG (area of influence, difference, filter, etc.) they will be discounting, from the initial surface of the study area, the surface occupied by the restrictions to produce a new thematic layer that will contain the feasible locations (Fig. 3).

The feasible locations occupy an area of 1,036.11 km^2. Their surface is composed of 17,740 cadastral parcels according to the Cadaster General Directorate and these constitute the alternatives under analysis in the later stages. The following steps consist of selecting, from among the above alternatives, which are the best to implement this type of facilities, based on a number of criteria.

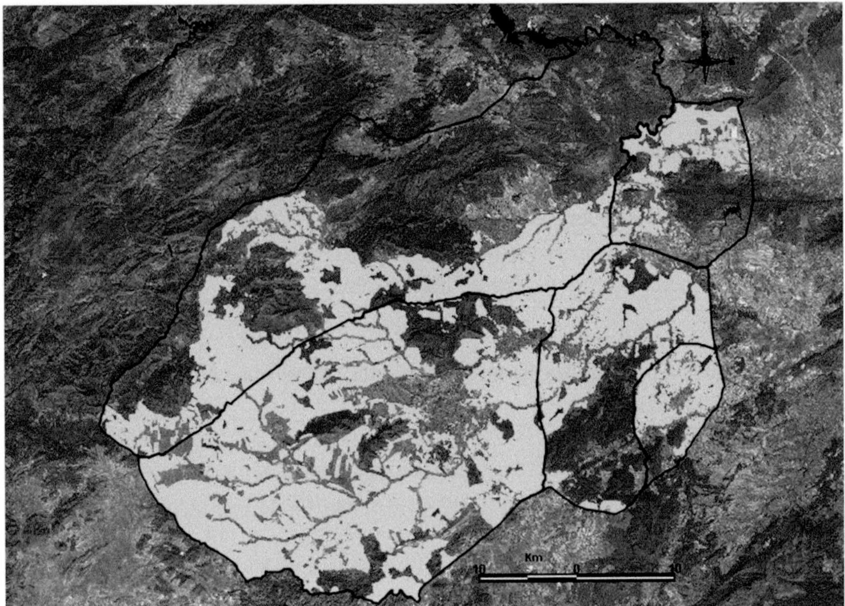

Fig. 3 Feasible locations to install photovoltaic solar farms

2.3 Stage 2: Optimal Locations Analysis

Criteria

In implementing renewable energy facilities, not only must it be taken into account that the analyzed zone is not affected by any legal restrictions, but we also have to rely on a series of criteria that influence the selection of the best location. Although research and studies have been conducted that define characteristics which these criteria must contain [16, 17], the choice of these criteria will depend mainly on the area of study. Therefore, following the guidelines indicated in Aran Carrion et al. [26], for the particular case, three groups of general criteria (location, orographic, and climatological) that are defined, will be broken down into a number of specific criteria which constitute the set of criteria which will influence the location, i.e., those that will opt for one location rather than another. Below we describe briefly each of the above-mentioned specific criteria:

C_1: Slope (%): Land slope, the higher percentage of having a surface inclination, the worse aptitude to hold a solar plant

C_2: Area (m^2): surface contained within a perimeter of land that can accommodate an RE installation

C_3: Field Orientation (degrees): Position or direction of the ground to a cardinal point. The most focused point is land oriented to the geographic South (270 °)

C_4: Distance to main roads (m): space or interval between the (Highway) road network and the different possible locations

C_5: Distance to power lines (m): space or interval between the nearest power line and the different possible locations

C_6: Distance to towns or villages (m): space or interval between centers of population and the different possible locations

C_7: Distance to electricity transformer substations (m): Space or interval between transformer substations of electric power and the different possible locations

C_8: Solar irradiation (kJ/m^2 day): value of the amount of solar irradiation that a field receives per unit area (m^2)

C_9: Average temperature (°C): corresponds with the average annual temperature in the different possible locations

Database

In order to create a database containing all alternatives and criteria, we followed a similar process to that followed in the definition of constraints. Thus, there were a total of 17,740 alternatives object of analysis, displayed by thematic information represented in rows and columns, so that rows constitute geographical objects which in this case will be alternatives to select linked (plots), and the columns will define named attributes or thematic variables (cadastral information and criteria) constituting an array with data relating to each plot for each of the nine criteria above. The attribute table is the database that will be used in the third stage.

2.4 *Stage 3: Selection of Optimal Locations*

The database created with gvSIG allows to obtain the numeric values of all the criteria for each of the alternatives. These values will be used to carry out a process of filtering according to a classification of alternatives by categories. To do so, an expert in photovoltaics will state not only the number of categories in which the alternatives based on the nine criteria should be classified, but he also establishes the limits of such categories for each of the criteria. According to the expert, it is possible to classify the alternatives into four categories (Cat 1, Cat 2, Cat 3, Cat 4) depending on the fitness or capacity for a solar farm (regular, good, very good, and excellent capacity, respectively), and the limits set (Table 2) of these categories based on the domains of the criteria that influence the decision.

Initially, the expert considers that the worst classification should only be regular (category 1) because, when determining in stage 1 areas in which a solar farm cannot be implanted, there is no alternative that can be termed as poor. Similarly, there should be a category defined as excellent (category 4) with the objective of being restrictive when making the selection process.

The selection process will consist in gradually, and using screening techniques, eliminating those choices that are lower, and thus obtain the sites located in the top

Table 2 Boundaries of categories of alternatives A_i for each of the criteria

Criteria	Cat 1	Cat 2	Cat 3	Cat 4
Slope (%)	Ai > 30	30 ≥ Ai > 20	20 ≥ Ai > 10	Ai ≤ 10
Area (m^2)	Ai < 1500	1500 ≤ Ai < 3500	3500 ≤ Ai < 10000	Ai ≥ 10000
Field orientation (°)	45 ≤ Ai < 135	135 ≤ Ai < 225	0 ≤ Ai < 45	225 ≤ Ai < 360
Distance to main roads (m)	Ai > 10000	10000 ≥ Ai > 5000	5000 ≥ Ai > 100	Ai ≤ 100
Distance to power lines (m)	Ai > 10000	10000 ≥ Ai > 3000	3000 ≥ Ai > 100	Ai ≤ 100
Distance to town or villages (m)	Ai < 100	100 ≤ Ai < 500	500 ≤ Ai < 1000	Ai ≥ 1000
Distance to electricity transformer substations (m)	Ai > 15000	15000 ≥ Ai > 10000	10000 ≥ Ai > 7500	Ai ≤ 7500
Solar irradiation (kJ/m^2 day)	Ai < 1200	1200 ≤ Ai < 1700	1700 ≤ Ai < 2000	Ai ≥ 2000
Average temperature (°C)	Ai < 12	12 ≤ Ai < 15	15 ≤ Ai < 17	Ai ≥ 17

categories. Therefore, the first step will consist of deleting those alternatives that have values in some of their criteria in category 1 in order to reduce the number of alternatives to those whose capacities of reception are good, very good or excellent (categories 2, 3 and 4, respectively). Once this first filtering has been completed, a new thematic layer (Fig. 4a) will have reduced the number of alternatives, so from

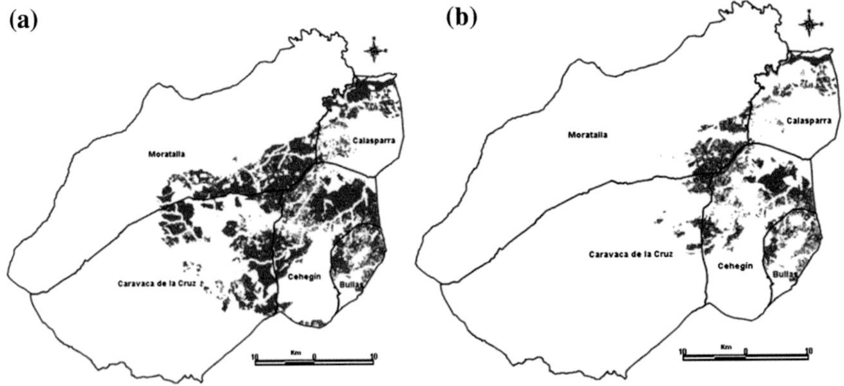

Fig. 4 **a**, **b** Alternatives resulting after removing those alternatives with criteria in category 1 (filtering n°1) and category 2 (filtering n°2)

17,740 possible initial locations it this case they will have been reduced to 8,961, all placed in categories 2, 3 and 4.

Continuing with the process of selection, the previous thematic layer (Fig. 4a) will be taken as a starting point and out a new filter will be carried out that removes those alternatives with values in some of their criteria in category 2. By so doing, the 8,961 will be reduced to 3,496 alternatives (Fig. 4b), all in categories 3 and 4.

Once this second filtering has been done, we will proceed analogously performing a third filter in order to obtain the best alternatives; that is those with all their criteria situated in the best category (Category 4), reapplying the gvSIG filter command to the 3,496 alternative obtained in the second filtering it is reduced to only seven. The location and identification of these alternatives are shown in Fig. 5.

Figure 5 shows that there is no optimal alternative located in the municipalities of Moratalla, Cehegín and Bullas. Most of the optimal alternatives are located in the municipality of Calasparra (specifically six of the seven best alternatives are located in it) and the remaining optimal alternative is located in the municipality of Caravaca de la Cruz.

Analyzing the criteria values for optimal alternatives it is observed that the criteria C_4 (Distance to main roads), C_5 (Distance to power lines) and C_9 (Average temperature) have the same values. Therefore they have no influence on the choice of the best alternative. It is also noticeable (Table 3) that certain criteria such as slope C_1, field orientation C_3 and solar irradiation C_8 have very similar values to

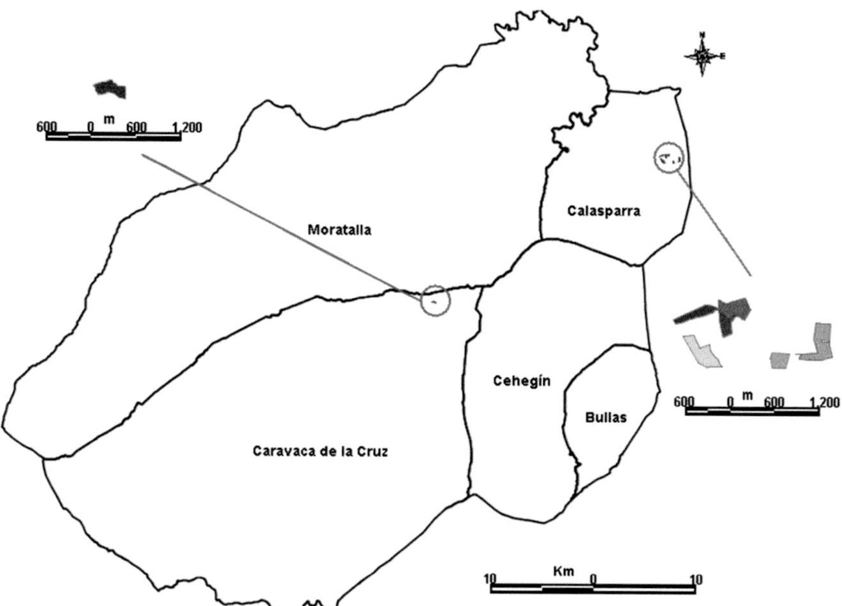

Fig. 5 Alternatives resulting after removing those alternatives with criteria in category 3 (filtering nº3)

Table 3 Values of criteria for optimal alternatives

Alternatives	C_1	C_2	C_3	C_6	C_7	C_8
	Min.	Max.	Max.	Max.	Min.	Max.
A_1	0.26	89298.37	360.00	6028.68	5793.66	2086
A_2	0.34	43903.39	360.00	6400.45	6192.66	2081
A_3	0.25	76262.28	360.00	6376.42	6174.85	2081
A_4	0.29	44092.23	360.00	7225.17	6989.63	2062
A_5	0.48	106290.92	360.00	7547.42	7319.09	2070
A_6	0.77	61831.30	359,57	1120.08	2002.35	2031
A_7	0.30	54748.54	360.00	5847.85	5624.50	2084

each other which means that these values are not very influential in the decision. Only criteria such as the area of the plot C_2, Distance to town or villages C_6 and distance to electricity transformer substations C_7 have variable values, therefore choosing the best location is determined by the weight or importance coefficient of these last criteria.

3 Conclusions

With this study it was found that the GIS software are not only excellent tools able to solve and visualize complex location problems, but also that they can generate important databases which provide an ideal starting point to address any problems of territorial nature.

In the proposed particular case different conclusions have been reached: In relation to obtaining suitable surfaces for locating photovoltaic solar farms (Fig. 3), it is concluded that the Northwest Region of Murcia is an optimal place to implement such facilities because, once all the restrictions have been considered, we have obtained a high percentage of suitable area available (43.54 %).

With the tools of GIS software and using the information provided by experts, it has been possible to perform a search and selection of the best places to locate such facilities, successfully reducing the initial alternatives to only a very small and manageable number of alternatives (Fig. 5).

Among the limitations of this study which could be included in possible future work one might mention extending the case study to the whole national territory or other areas where there is a desire to implement solar farms, as well as to increase the number of renewable technologies to be implemented (wind, solar thermal, biomass, biogas, etc.). It would also be interesting to combine GIS with other decision support tools such as multicriteria decision methods with the aim of establishing a comparison of methodologies for evaluating the different locations available.

Acknowledgment María Teresa Lamata and José Luis Verdegay want to acknowledge Christer Carlsson for his support, help and sincere friendship along the last 25 years. This work has been partially supported by FEDER funds, the DGICYT and Junta de Andalucía under projects TIN2014-55024-P and P11-TIC-8001, respectively.

References

1. Arrhenius, S.: On the influence of carbonic acid in the air upon the temperature of the ground. Philos. Mag. J. Sci. **5**(41), 237–276 (1896)
2. United States.: Energy information administration. In: International Energy Outlook 2013. With Projections to 2040. Office of Energy Analysis U.S. Department of Energy Washington, DC 20585, pp. 1–312 (2013)
3. United Nations.: Report of the united nations. In: Conference on environment and development. Rio Declaration on Environment and Development. Rio de Janeiro (1992)
4. United Nations: Framework convention on climatic change: Report of the conference of the parties on its third session. Adoption of the Kyoto Protocol, Kyoto (1997)
5. Working Group I: Climate change. In: Houghton, J.T., Jenkins, G.J., Ephraums, J.J. (eds.) The IPCC Scientific Assessment. Cambridge University Press, Cambridge (1990)
6. Working Group II: Climate change. In: Tegart, W.J.McG., Sheldon, G.W., Griffiths, D.C. (eds.) The IPCC Impacts Assessment. Australian Government Publishing Service, Camberra (1990)
7. Working Group III: Climate change. The IPCC Response Strategies. In: World Meteorological Organization/United Nations Environment Program. Island Press, Covelo (1990)
8. European Commission.: Energy for the Future: Renewable Sources of Energy—Green Paper for a Community Strategy, Brussels (1996)
9. Commission, European: Energy for the future: renewable sources of energy. White Paper for a Community Strategy and Action Plan, Brussels (1997)
10. Dovì, V.G., Friedler, F., Huisingh, D., Klemes, J.J.: Cleaner energy for sustainable future. J. Clean. Prod. **17**, 889–895 (2009)
11. Institute for Energy Diversification and Saving IDAE.: Renewable Energies Plan (PER) 2005–2010. Ministry of Industry, Tourism and Commerce, Madrid (2005) (in Spanish)
12. Institute for Energy Diversification and Saving IDAE.: Renewable Energies Plan (PANER) 2011–2020. Ministry of Industry, Tourism and Commerce, Madrid (2010) (in Spanish)
13. European Commission.: PV Status Report 2010: Research, Solar Cell Production and Market Implementation of Photovoltaics. DG Joint Research Centre. Institute for Energy, Renewable Energy Unit, ISBN 978-92-79-15657-1, doi:10.2788/87966, EUR 24344 EN—2010. Italy (2010)
14. Urbina, A.: Solar electricity in a changing environment: the case of Spain. Renewable Energy **68**, 264–269 (2014)
15. Gómez-López, M.D., García-Cascales, M.S., Ruiz-Delgado, E.: Situations and problems of renewable energy in the region of Murcia, Spain. Renew. Sustain. Energy Rev. **14**, 1253–1262 (2010)
16. Janke, J.R.: Multicriteria GIS modeling of wind and solar farms in Colorado. Renewable Energy **35**, 2228–2234 (2010)
17. Uyan, M.: GIS-based solar farms site selection using analytic hierarchy process (AHP) in Karapinar region, Konya/Turkey. Renew. Sustain. Energy Rev. **28**, 11–17 (2013)
18. Legislative Decree 1/2005 of 10 June, in which the revised text of the land law in the Region of Murcia is approved. BORM No. 282. Governing Council of the Autonomous Community of the Region of Murcia. Murcia (2005) (in Spanish)
19. Decree 102/2006 of 8 June, in which the guidelines and land use plan of industrial land in the Region of Murcia are adopted. BORM No. 137, Industry and Environment Counselling,

General Management Planning and coast, Promotion Institution of the Region of Murcia. Murcia (2006) (in Spanish)
20. Law 42/2007 of 13 December, on the Natural Heritage and Biodiversity. BOE no 299. Ministry of Agriculture, Food and Environment. Madrid (2007) (in Spanish)
21. Law 4/2009 of 14 May, on Integrated Environmental Protection of the Region of Murcia. BORM no 116. Governing council of the Autonomous Community of the Region of Murcia. Murcia (2009) (in Spanish)
22. Law 3/1995, of 23 March, on cattle trails. BOE No. 71, Ministry of Agriculture, Food and Environment. Madrid (1995) (in Spanish)
23. Law 16/1985 of 25 June of Spanish Heritage. BOE No. 24 Ministry of Education, Culture and Sport. Madrid (1985) (in Spanish)
24. Law 25/1988 of 29 July, of Roads. BOE No. 182 Ministry for Building. Madrid (1998) (in Spanish)
25. European Parliament.: Directive 92/43/EEC of 21 May 1992 on the conservation of natural habitats and of wild fauna and flora. Official Journal of the European Union. Brussels (2009)
26. Arán-Carrión, J., Espín-Estrella, A., Aznar-Dols, F., Zamorano-Toro, M., Rodríguez, M., Ramos-Ridao, A.: Environmental decision-support systems for evaluating the carrying capacity of land areas: optimal site selection for grid-connected photovoltaic power plants. Renew. Sustain. Energy Rev. **12**, 2358–2380 (2008)

Author Index

M. Collan et al. (eds.), *Fuzzy Technology*, Studies in Fuzziness and Soft Computing 335, DOI 10.1007/978-3-319-26986-3